Polyvinyl Alcohol/Halloysite Nanotube Bionanocomposites as Biodegradable Packaging Materials

Zainab Waheed Abdullah · Yu Dong

Polyvinyl Alcohol/Halloysite Nanotube Bionanocomposites as Biodegradable Packaging Materials

 Springer

Zainab Waheed Abdullah
School of Civil and Mechanical Engineering
Curtin University
Perth, WA, Australia

Yu Dong
School of Civil and Mechanical Engineering
Curtin University
Perth, WA, Australia

ISBN 978-981-15-7358-3   ISBN 978-981-15-7356-9   (eBook)
https://doi.org/10.1007/978-981-15-7356-9

This Springer imprint is published by the registered company Springer Nature Singapore Pte Ltd.
The registered company address is: 152 Beach Road, #21-01/04 Gateway East, Singapore 189721, Singapore

Preface

The accumulation of plastic wastes is one of the urgent environmental concerns around the world. Petroleum-based polymers have been widely used in many sectors for a wide range of daily-life applications such as constructions, appliances, medical devices and food packaging to replace conventional metals, wood and ceramics due to their better mechanical, thermal and barrier properties in light-weight structures. Nonetheless, it is worth mentioning that non-degradable material characteristics for such polymers mean the lack of environmental sustainability, recyclability and renewability despite their impressive material properties. In particular, large portions of used petroleum-based polymers become solid wastes that can be burned at the end of life to eliminate their accumulation in the simplest way moving forward leading to higher carbon emission and global warming. As such, the implementation of ecofriendly and biodegradable polymers as alternative resources is urged in the wide communities to echo "sustainable cities and communities" and "climate action" among 17 Sustainable Development Goals proposed by United Nations. In particular, current concentration has been placed on single-use area of bioplastics/biodegradable polymers in short-life food packaging applications.

Biodegradable polymers (or biopolymers) are regarded as one of the future advanced materials to ultimately eradicate the crisis of plastic wastes. This is because biopolymers, derived from natural resources like plants, animals and microorganisms, have the great capability to degrade at much faster pace than petroleum-based polymers in natural environments like soil and sludge. However, weak properties of such biopolymers can limit their potential industrial applications if only used as neat materials. Hence, mixing with other biopolymers as blends or reinforced with different additives particularly on a nanoscale level as bio-nanocomposites can yield their highly improved properties and broadened application fields.

This book holistically covers the development, characterisation and modelling of novel polyvinyl alcohol (PVA)/starch (ST)/glycerol (GL)/halloysite nanotube (HNT) bionanocomposite films as potential sustainable and biodegradable materials for food packaging applications. PVA is a popular water-soluble synthetic polymer with good tensile strength, Young's modulus and gas barrier properties. Moreover, its drawbacks can be reflected from low biodegradability, small elongation at break, poor water resistance and high water solubility. Consequently, it has been blended with ST and GL to improve biodegradation rates and elongation at break, respectively, as well as these blends are reinforced with HNTs to enhance water resistance and water/gas barrier properties. This book has been developed from the Ph.D. thesis of Zainab Waheed Abdullah as the first author comprising totally 6 chapters. Chapter 1 provides a comprehensive review of most popular PVA blends and their corresponding bionanocomposites in terms of manufacturing processes, material properties and key applications mainly used in food packaging. Chapter 2 introduces the raw materials, manufacturing procedures and conditions as well as sophisticated material characterisation techniques. Chapter 3 discusses the effect of material composition and nanofiller content on mechanical, thermal and optical properties along with morphological structures of neat PVA, PVA blends and PVA/ST/GL/HNT bionanocomposites. Chapter 4 presents the water resistance and biodegradability of such bionanocomposites in comparison with those of neat PVA and PVA blends. Chapter 5 focuses on water barrier properties and gas permeabilities of PVA, PVA blends and PVA/ST/GL/HNT bionanocomposites. Finally, Chap. 6 describes the overall migration rates of PVA/ST/GL blend films and their bionanocomposites as well as the nanoparticle migration rates of HNTs along with the detailed food packaging tests using peaches and freshly cut avocados.

The first author Zainab Waheed Abdullah acknowledges the Higher Committee for Education Development (HCED) in Iraq to offer the Ph.D. scholarship at Curtin University, Perth, Australia, to support her research project about biodegradable packaging bionanocomposites where this book is based upon. Special thanks are also indebted to Graeme Watson, David Collier and Andy Viereckl at Mechanical Engineering Laboratory, Elaine Millers at Microscopy and Microanalysis Facility in the John de Laeter Centre, Dr. Thomas Becker at School of Molecular and Life Sciences, and Andrew Chan, Ning Han and Prof. Shaomin Liu at WA School of Mines: Minerals, Energy and Chemical Engineering, Curtin University, Australia, for their continued technical supports. Finally, the editorial staff at Springer Nature particularly Loyola D'Silva and Mohan Mathiazhagan for kind editorial assistance in successfully publishing this book.

Last but not least, any comments and feedbacks from peers in academia and industries are welcome to improve overall book quality and shine upon the emerging knowledge of bionanocomposite materials for sustainable and biodegradable food packaging.

Perth, Australia   Zainab Waheed Abdullah
                                                                              Yu Dong

Contents

Chapter 1
Introduction

Abstract Environmental pollution due to solid plastic wastes has drawn great attention over decades in academia and industries. Using petroleum-based (non-degradable) polymers in several daily applications, particularly for food packaging, has even more severe impact on environmental sustainability. Only a small portion of such solid plastic wastes can be recycled, while the majority is accumulated permanently in the environment. As such, it is essential to replace petroleum-based polymers with other biodegradable polymers. Most of these biodegradable polymers have limited mechanical, thermal and barrier properties leading to their narrow applications as neat polymers. Consequently, it is motivated by the material development of new biopolymer blends and bionanocomposites to meet stringent requirements of these applications particularly for food packaging. Polyvinyl alcohol (PVA) is one of these biodegradable polymers, widely used as a neat polymer or blend matrices for bionanocomposites. This chapter reviews in detail the most popular PVA blends and bionanocomposites with respect to their manufacturing processes, properties and key applications mainly targeting food packaging.

Keywords Biodegradable polymers · Polyvinyl alcohol (PVA) · PVA blends · PVA bionanocomposites

Polyvinyl alcohol (PVA) is a synthetic water-soluble biopolymer. It has good mechanical and thermal properties as well as relatively high cost and limited biodegradability in some environments like soil [1]. As a typical water-soluble polymer, PVA has many hydroxyl groups leading to a decrease in water vapour permeability (WVP) [2]. On the other hand, it has low gas permeability and good biocompatibility as well as non-toxicity for increasing its applications in medical sectors such as contact lens, eye drops and tissue adhesion barriers and packaging sectors like food packaging [3–5]. PVA is blended successfully with other biopolymers such as starch (ST) to improve its biodegradability [6], chitosan for better antimicrobial properties [7], polylactic acid (PLA) to improve mechanical and thermal properties [8] to mention a few. On the other hand, barrier properties of PVA can be promoted significantly with the incorporation of nanofillers [9].

© Springer Nature Singapore Pte Ltd. 2020
Z. W. Abdullah and Y. Dong, *Polyvinyl Alcohol/Halloysite Nanotube Bionanocomposites as Biodegradable Packaging Materials*,
https://doi.org/10.1007/978-981-15-7356-9_1

1.1 Biodegradable Polymers

Biopolymers, also known as biodegradable, biocompatible, environmentally friendly, sustainable, renewable and green polymers, have been developed rapidly in several industrial sectors particularly in material packaging in the last decade [10]. The global market of biopolymers reached 103,000 tonnes in 2016 with a further expansion expected to reach 884,000 tons by 2020 [11, 12]. Biopolymers are produced from alternative resources such as the direct extraction from biomass and chemical synthesis from biomass as well as the production from microbial resources [13, 14]. Although these biopolymers are widely available, relatively cheap, non-toxic, biocompatible and biodegradable with high reactivity and acceptable strength, they still have narrow applications due to weak thermal stability and poor barrier properties [5, 13, 15]. The most popular way to overcome these limitations is developing polymer nanocomposite systems [16, 17]. Nanotechnology has become a major focus for packaging applications due to the development of new material systems, which have unique properties resulting from the incorporation of nanoscaled materials with large surface-area-to-volume ratios. This leads to enhanced thermal and chemical stabilities, as well as better mechanical and barrier properties with lower density, as opposed to micro- and macro-scaled materials [18]. According to Youssef and El-Sayed [19] and Duncan [20], the association of nanotechnology in food packaging applications was worth US$4.13 billion in 2008 and increased to US$7.3 billion in 2014 with an annual increase up to 12%, while nanotechnology is expected to reach US$3 trillion in 2020 across the global economy.

1.1.1 Concept of Biodegradability

Biodegradation can be defined as a natural decomposition process of organic materials, which breaks down to simple components by an enzymatic action of microorganisms such as bacteria and fungi in appropriate environmental conditions [21–23]. This process is mostly associated with the loss of mechanical properties and change of chemical characteristics [17]. New simple biomass, water and CO_2 are end products of this process in the presence of oxygen (aerobic conditions) or methane without oxygen (anaerobic conditions) [22, 24]. Furthermore, biodegradable materials like biodegradable polymers can be validated according to ASTM-D20-96, ISO-472, EN-13432-2000 and DIN-103.2 standards since such materials can suffer from bonding scission in the backbone leading to significant changes in their chemical and morphological structures under particular environmental conditions and by the action of microorganisms [22, 23]. Such environmental conditions include temperature, relative humidity, available microorganism types and numbers. In other words, biodegradation is much faster at the temperature range of 50–70 °C when compared with that at room temperature. It is also the case for relative humidity because biodegradation becomes faster in humid soil than dry counterpart [13, 25]. In addition to environmental conditions, biodegradation can also be controlled by other material-related factors like molecular weight, length of polymeric chains, crystallinity degree, treatment history, functional groups and additives like plasticisers [17, 21].

Biodegradation process involves multi-steps as follows [21, 24]:

- **Biodeterioration**: Biodegradable polymers are fragmented into very small fractions by the action of microorganisms.
- **Depolymerisation**: Small fractions of polymers are cleaved by microorganisms with a significant reduction in molecular weight to produce monomers, oligomers and dimers.
- Some of the molecules are produced in the step of depolymerisation, which are consumed by microorganism cells and go across the plasmic membranes, while the rest of molecules may stay in the extracellular surroundings subjected to different modifications.
- **Assimilation**: Consumed molecules are integrated into the metabolism of microorganisms to produce energy, storage vesicles and new biomass.
- **Mineralisation**: CO_2, N_2, CH_4, H_2O and different salts are finally released from intracellular metabolites to the environment after complete oxidisation [21, 24].

The biodegradation of polymers can be determined using several standard testing methods [21, 24]:

- **Visual changes**: Biodegradable polymers show visible changes like increased surface roughness, colour changes, fragmentation, crack and hole formation. These changes can be indicative of typical biodegradation but cannot be used as an evaluation method. Scanning electron microscopy (SEM) and/or atomic force microscopy (AFM) are generally employed to obtain more informative details relating to biodegradation mechanism and steps.
- **Measurement of weight loss**: Biodegradable polymers have their mass loss as the degradation is in progress. Therefore, weight loss or residual polymer weight is employed to evaluate the biodegradation rate. Thorough cleaning of polymeric samples is a crucial step during this process particularly with the soil and composting biodegradation.
- **Variation of material properties**: When polymer samples show weight loss, their rheological properties are also altered accordingly. Mechanical properties, particularly tensile strength, can be highly sensitive to weight change causing a reduction in material thickness, as well as thermal properties like glass transition temperature (T_g) and degree of crystallinity (X_c) that are significantly altered as an indicator of material degradation.
- **Product formation**: CO_2 is the end product in a biodegradation process under aerobic conditions by the consumption of O_2 to obtain oxidised carbon in polymeric samples. Hence, the evaluation of O_2 consumption and CO_2 production is another signal of biodegradation, as well as the generation of other end products like glucose and acids [21, 24].

1.1.2 Types of Biodegradable Polymers

Biopolymers can be classified according to their resources and synthetic processes as follows [13, 14, 22, 26]:

- Biopolymers are extracted directly from biomass resources including plant carbohydrate (e.g. starch, cellulose, agar, etc.), plant and animal protein (e.g. soy protein, collagen, gelatin, etc.), and plant and animal lipids (e.g. wax and fatty acids).
- Biopolymers are chemically synthesised from biomass resources (e.g. PLA) and from petrochemical resources (e.g. PVA, poly(glycolic acid) (PGA), poly (ε-caprolactone) (PCL), etc.).
- Biopolymers are produced by microbial fermentation such as microbial polyesters (e.g. poly(hydroxyalkanoates) (PHAs), poly(β-hydroxybutyrate) (PHB), etc.) and microbial polysaccharides (e.g. curdlan and pullulan) [13, 14, 22, 26], as shown in Fig. 1.1.

Most of these neat biopolymers have narrow applications because of their poor thermal and barrier properties, as well as limited processability despite their good biodegradability, biocompatibility, non-toxicity and viability. Consequently, blending biopolymers with other petroleum-based polymers and reinforcing them with nanofillers is a feasible solution to overcoming these limitations [13, 27].

1.2   PVA and PVA Blends

PVA is a water-soluble polymer with hydrocarbon backbone [2, 4]. It was synthesised first in 1924 by Herrmann and Haehnel by means of the hydrolysis of polyvinyl acetate (PVAc) with potassium hydroxide in ethanol [1, 28]. PVA is not produced from the direct polymerisation of corresponding monomers. It is manufactured today from the parent homopolymer PVAc [1, 2, 29]. Vinyl acetate is polymerised in the presence of alcohol solution like methanol or ethanol via a free-radical mechanism to produce PVAc, and PVA is synthesised by hydrolysing PVAc in one-pot reactor [2, 30]. Depending on the hydrolysis degree (HD) of PVAc, different grades of PVA can be produced in a wide range of molecular weight from 20×10^3 to 400×10^3 g/mol and HD levels of 70–99%. PVA is a semicrystalline polymer consisting of 1,3-diol or 1,2-diol units relative to the HD of PVAc with many hydroxyl groups on the surfaces [4, 31] according to Fig. 1.2.

Molecular weight and HD determine most PVA properties, as stated in Fig. 1.3. For example, tensile strength, adhesion strength, water and solvent resistance of PVA increase with increasing the molecular weight and HD while the solubility, water sensitivity and flexibility decrease in an opposite trend [3, 4, 32].

Full-hydrolysis PVA has limited ductility, and its melting temperature (T_m) is very close to the decomposition counterpart. As a result, it is not considered as a

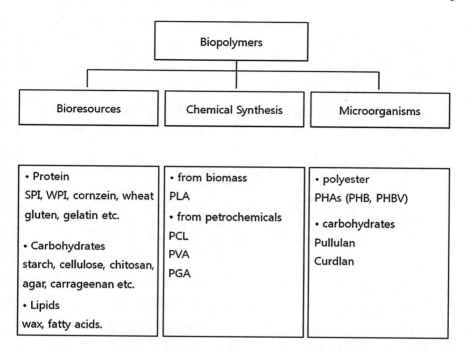

Fig. 1.1 Classification of biopolymers. Image taken from [13] with the copyright permission from Elsevier

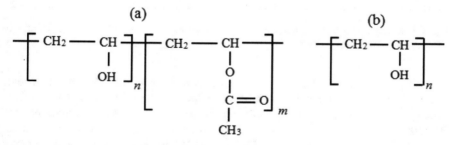

Fig. 1.2 Schematic diagram of PVA chemical structures: **a** partial-hydrolysis PVA and **b** full-hydrolysis PVA. Image taken from [4]

thermoplastic material without plasticisers [8, 30, 33]. Plasticisers can be defined as a low-molecular-weight and non-volatile organic compound in possession of a high boiling temperature without the separation from the blends, while the T_m and T_g are reduced for polymers to improve their material processability and flexibility [34, 35]. Consequently, the addition of plasticisers is essential to diminish these limitations particularly during a thermal process like blow moulding and screw extrusion widely employed for material packaging applications [8]. PVA starts the

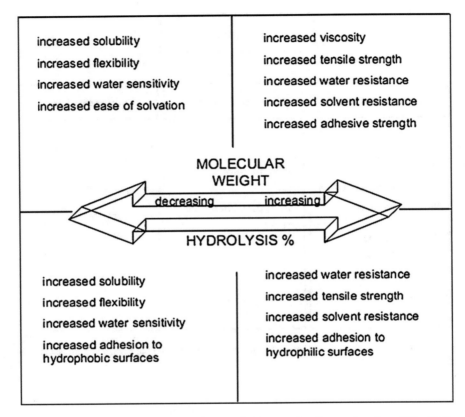

Fig. 1.3 Effects of molecular weight and HD on PVA properties. Image taken from [32] with the copyright permission from Elsevier

thermal decomposition around 150 °C, during which many water molecules would release from polymeric molecules to be recovered with the addition of water and/or organic plasticisers [2]. Furthermore, the presence of plasticisers improves the flexibility and reduces the shrinkage during the processing, handling and storage steps [8]. On the other hand, partial-hydrolysis PVA is a copolymer of vinyl acetate and vinyl alcohol due to the presence of some residual acetate groups [36]. These groups diminish the formation of hydrogen bonding with adjacent hydroxyl groups in PVA blends leading to a decrease in water resistance of blends owing to many free hydroxyl groups. Consequently, partial-hydrolysis PVA is not a suitable material selection for specific applications with the requirement of high water resistance like material packaging [37, 38].

PVA has mixed material properties like good transparency, non-toxicity, odourlessness, good compatibility, relatively high biodegradability in some environments and high mechanical properties despite its poor barrier properties, limited thermal stability and relatively high cost when compared with petroleum-based polymers. In order to overcome these limitations by blending PVA with other polymers and/or incorporating nanofillers, PVA is an ideal base material for many different applications particularly in biomedical devices and food packaging [1, 3, 4]. As such, in the following sections, the most popular PVA blends and nanocomposite systems are discussed in detail.

1.2.1 PVA/ST Blends

Starch (ST) is a completely biodegradable, biocompatible and renewable polymer belonging to polysaccharide family. It is a naturally available polymer with relatively good transparency, non-toxicity, odourlessness, tastelessness and cost-effectiveness [39]. ST has a chemical formula $C_6H_{10}O_5$ consisting of two different biomacromolecules that are known as amylose and amylopectin as well as minor contents of protein, phosphor and lipids [40, 41]. Amylose has linear biomacromolecules in which D-glucose units are linked by $\alpha(1$–$4)$ linkage with the molecular weight of 10^5–10^6 g/mol. Whereas, amylopectin has multi-branched biomacromolecules where D-glucose units are linked by $\alpha(1$–$4)$ at the backbone and $\alpha(1$–$6)$ at the branches with a relatively high molecular weight of 10^7–10^9 g/mol when compared with amylose counterpart [39, 40, 42], as illustrated in Fig. 1.4.

ST is a semicrystalline polymer with its crystallinity degree ranging from 15 to 45% depending on amylose and amylopectin contents [43]. The approximate contents of these two macromolecular components are in range of 15–30% for amylose and 70–80% for amylopectin depending on ST resources [39, 44], as summarised in Table 1.1.

ST with many hydrogen bonds between molecules can restrict the mobility of polymeric chains, reduce the flexibility and increase the brittleness [43]. Furthermore, the T_m of ST at 220–224 °C is close to its decomposition temperature at 220 °C, which makes it harder for material processability as a neat polymer. Consequently, thermoplastic starch (TPS) generated in a gelatinisation process is used widely instead of neat ST [43]. Gelatinisation process is defined as a process to destroy highly organised crystalline phase of ST and convert it to amorphous phase in order to produce a plasticised ST paste in the presence of heat, water and/or plasticisers. In this process, hydrogen bonds between ST molecules are replaced with other bonds between the molecules of ST and plasticisers, leading to the improvement of flexibility and the reduction in T_m [44, 46]. TPS still has limitations such as high water sensitivity, poor mechanical strength, and limited dimensional and thermal stabilities [47, 48]. Consequently, blending PVA with ST enhances mechanical and barrier properties of ST, as well as improves PVA biodegradability with the cost reduction [48, 49]. Tânase et al. [6] found that soil biodegradation of PVA/ST blends could be improved by 32.45% with increasing the ST content from 0 to 20 wt% when compared with that of neat PVA. PVA/ST blends have been widely studied since the 1980s for the film production via casting and calendaring

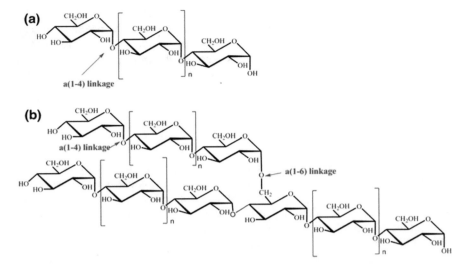

Fig. 1.4 Chemical structures of starch: **a** amylose and **b** amylopectin. Image taken from [39] with the copyright permission from Springer

Table 1.1 Amylose and amylopectin contents of starch corresponding to their resources

Starch	Amylose (%)	Amylopectin (%)	References
Maize	26–28	71–73	[44]
	25	75	[45]
Waxy maize	<1	>99	[39, 44]
	0	100	[45]
Amylomaize	48–77	23–52	[39]
	50–80	20–50	[44]
Amylomaize-5	53	47	[45]
Corn	17–25	5–83	[39]
High amylose corn	55–70	30–45	[39]
Amylomaize-7	70	30	[45]
Potato	17–24	76–83	[39]
	20–25	74–79	[44]
	22	78	[45]
Wheat	20–25	75–80	[39]
	26–27	72–73	[44]
	23	77	[45]
Rice	15–35	65–85	[39]
	19	81	[45]
Chickpeas	30–40	60–70	[39]

(continued)

Table 1.1 (continued)

Starch	Amylose (%)	Amylopectin (%)	References
Tapioca (cassava)	19–22	28–81	[39]
	17	83	[45]
Banana	17–24	76–83	[39]
	20	80	[45]
Cush-Cush Yam	9–15	85–91	[39]
Shoti	30	70	[45]

methods in order to replace polystyrene in material packaging [50]. These blends have high compatibility in the presence of plasticisers corresponding to chemical interactions between their hydroxyl groups to produce strong hydrogen bonds [32, 51–53], as illustrated in Fig. 1.5. Sin et al. [54] confirmed via differential scanning calorimetry (DSC) that PVA/ST blends were compatible to some extent and the addition of 25–35 wt% ST to PVA yielded strong bonding interactions between PVA and ST similar to those between neat PVA molecules.

Glycerol (GL) is the best solution candidate as a plasticiser for PVA/ST blends due to their close solubility parameters, as evidenced by 21.10, 22.5 and 23.4 MPa$^{1/2}$ for PVA, ST and GL, respectively [55]. Zanela et al. [56] prepared PVA/ST/GL blends with different blend ratios of 50:20:30, 30:40:30, 40:20:40, 45:20:35, 30:30:40 and 32.5:32.5:35 by weight and investigated the effect of each component on their material properties. All blend ratios reflected good processability with homogeneous morphological structures. Moreover, their results showed that a high PVA content improved mechanical and barrier properties, while a higher GL content reduced the mechanical strength and improved elongation at break due to its plasticisation effect. Furthermore, a much higher ST content improved biodegradability and reduced water resistance. Overall, the blend ratio is a crucial factor to determine the resulting properties.

1.2.2 PVA/PLA Blends

PLA is a biodegradable aliphatic polyester that is synthesised from biomass resources like potato, corn and cane sugar [57] with a chemical structure depicted in Fig. 1.6. Lactic acid monomers (L- and D-lactic acid) are obtained chemically or biologically from the fermentation of carbohydrates by lactic bacteria like *Lactobacillus* genius. Low-molecular-weight PLA can be produced from these monomers by the polycondensation reaction [22, 29], while high-molecular-weight PLA is obtained by means of open ring polymerisation (ORP) of lactide monomers [22, 58]. Different grades of PLA are commercially available depending on L-/D-lactic acid ratio such as PLLA (100% L-lactic acid) and PDLLA (copolymer of D, L-lactic acid) [22].

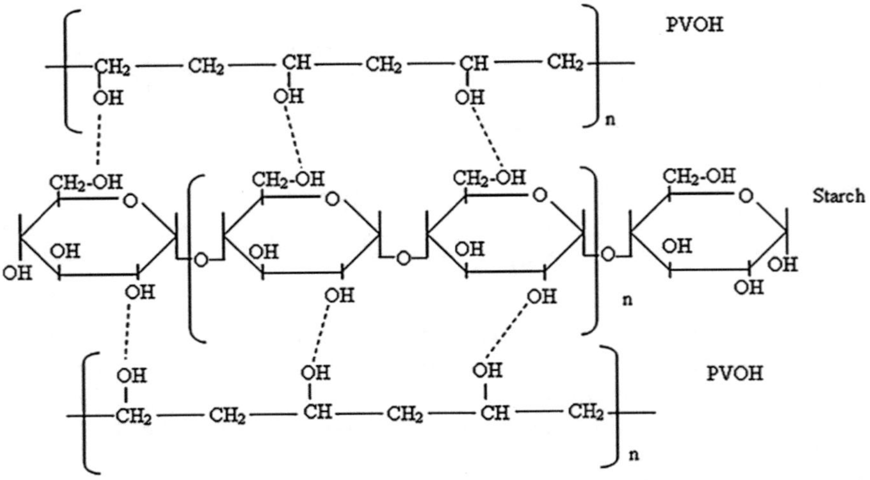

Fig. 1.5 Hydrogen bonds between PVA and ST (marked with dashed lines). Image taken from [32] with the copyright permission from Elsevier. Note that PVOH in the chemical structures denotes PVA

Fig. 1.6 Chemical structure of PLA. Image taken from [22] with the copyright permission from Springer

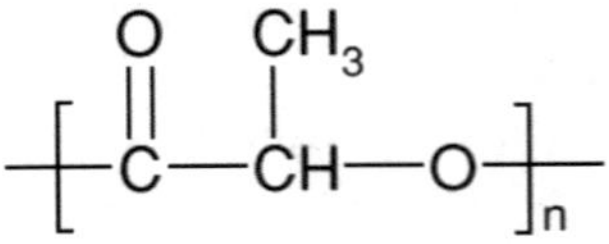

Although PLA has good mechanical, thermal and barrier properties, as well as biodegradability [29], its limited applications are quite evident due to low flexibility and high hydrophobicity leading to poor water uptake and slow hydrolytic degradation rates [59]. These limitations can be overcome by blending PLA with other biopolymers and petroleum-based polymers [31]. PVA is one of these polymers that can be blended widely with PLA to improve mechanical and thermal properties, as well as water resistance of PVA [60]. According to Shuai et al. [61], PVA/PLA blends had good miscibility due to the existence of single peaks for T_g and T_m in DSC curves. Similarly, Restrepo et al. [62] and Yeh et al. [63] found that single T_g was good evidence of the formation of a compatible binary system for PVA/PLA blends because of the interaction between hydroxyl groups of PVA with carbonyl groups of PLA. Moreover, PVA/PLA blends have better mechanical properties compared with neat polymers [8]. In addition, Hu et al. [64] prepared composite films by mixing PVA with ST/lactic acid graft (ST-g-PLA) copolymers. Their results showed the tensile strength and elongation at break for ST-g-PLA/PVA films were increased by 69.15 and 84.39%, respectively, as well as the water absorption was decreased by 50.39% when compared with those of ST/PVA films.

1.2.3 PVA/Chitin Blends

Chitin is a biodegradable and biocompatible polymer extracted from the shells of insects, crabs, shrimps and lobsters through the decalcification (i.e. acidic treatment), deproteination (i.e. alkaline treatment) and finally decolourisation for industrial-scaled production [65]. Chitin consists of β-(1→4)-2-acetamido-2-deoxy-D-glucopyranose units with small amounts of β-(1→4)-2-amino-2-deoxy-D-glucopyranose residues [22, 23, 65], as shown in Fig. 1.7. It is a highly acetylated biopolymer and widely available in nature as the secondary to cellulose [22]. When the acetylation degree of chitin is less than 50%, it would be known as chitosan [65]. This polymer has antimicrobial activities and is insoluble in water, but it also has the ability of water retention and moistening properties. As such, it is mainly applicable for packaging and cosmetic sectors [23].

Chitin has a highly ordered crystalline structure with a number of intra- and intermolecular hydrogen bonds [22]. According to Aoi et al. [66], such rigid crystalline structures are the main reason for poor water solubility and high intractability. Consequently, blending chitin with other biocompatible polymers like PVA can diminish these limitations [66]. Their DSC results showed that PVA/chitin blends had a single T_g due to high miscibility between components, which decreased with increasing the chitin content from 0 to 70 wt%. Furthermore, the biodegradability of PVA blends is relatively high in the presence of chitin because chitin is a completely biodegradable polymer [67].

1.2.4 PVA/Chitosan Blends

Chitosan is a linear aliphatic polyamide copolymer obtained mainly from the deacetylation of chitin, which is available in seafood crusts with a molecular structure of β(1,4)-2-amino-2-deoxy-D-glucose [29, 65, 68], as shown in Fig. 1.8. The cationic amino groups distribute around the chitosan's backbone resulting in antimicrobial properties against fungi, yeast and bacteria [65]. Chitosan is a semicrystalline polymer and the degree of crystallinity, and many other properties like viscosity and solubility are controlled by the deacetylation degree and molecular weight [22, 69]. The molecular weight and deacetylation degree are in range of 5×10^3–1×10^6 g/mol and 2–60%, respectively, depending on the source of chitosan [22].

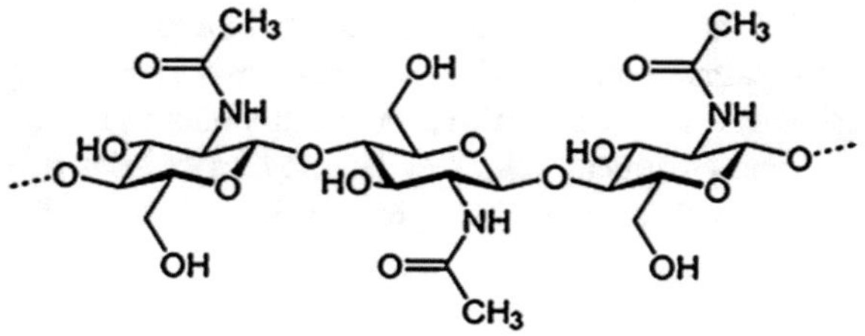

Fig. 1.7 Chemical structure of chitin. Image taken from [65] with the copyright permission from Elsevier

Chitosan is a biodegradable polymer in polysaccharide family with non-toxicity, wide availability, cost effectiveness and insolubility in water and most organic solvents, but it can be dissolved in acidic solutions with a pH value less than 6.3 [27, 65]. Furthermore, it has excellent processability and good barrier properties against gas and aroma despite limited mechanical strength [70, 71]. Consequently, blending hydrophilic PVA with biologically active chitosan could produce bene-ficial antimicrobial films with acceptable mechanical and barrier properties for many applications like food packaging with the shelf life improvement [7, 71]. Liu et al. [72] reported that PVA/chitosan blends at a weight blend ratio of 70:30 demonstrated smooth and homogenous surfaces on SEM morphology without typical defects such as phase separation, pores, bubbles and cracks, which was considered as an evidence of good compatibility between components. Nevertheless, increasing the chitosan content up to 35 wt% in PVA/chitosan blends showed high surface roughness. Similarly, He and Xiong [7] found PVA/chitosan blends had homogenous morphological structures with the aid of SEM due to the good compatibility between components. Furthermore, the swelling degree of PVA/ chitosan blends at a blend ratio of 2:3 in distilled water could be reduced by 60% when compared with that of neat PVA owing to an overall balance between hydroxyl groups of PVA and amino groups of chitosan. This led to a built-up rigid structure with the minimum number of free hydroxyl groups interacting with water molecules. Moreover, Tripathi et al. [71] used PVA/chitosan blends as a coating solution for fresh tomatoes, and their results showed a clear decline of fungi growth rate as opposed to uncoated counterparts, as well as increasing shelf life.

1.2.5 PVA/Gelatin Blends

Gelatin is a completely biodegradable, biocompatible and water-soluble polymer [23]. It is prepared mainly from collagen extracted from fibrous tissues of skins, bones, blood vessels and intervertebral disc [22, 73]. Depending on different pre-treatment methods of collagen, two types of gelatin are produced. Type A gelatin is derived by using acidic treatment, and type B gelatin is obtained from alkaline treatment [29, 73]. Both types consist of approximately 19 amino acid groups that are joined by peptide linkages with a typical structure of –Ala–Gly–

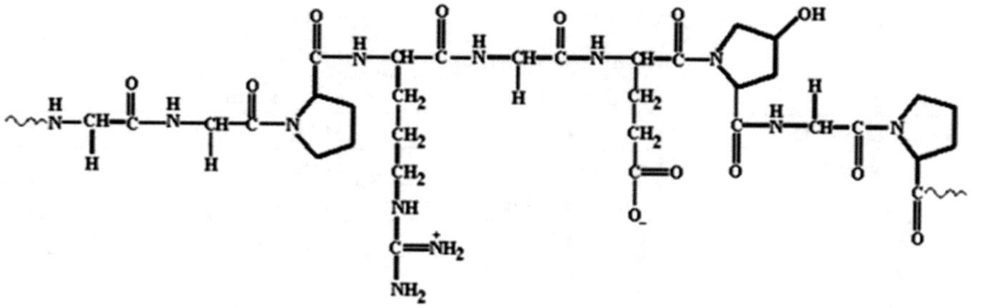

Fig. 1.9 Chemical structure of gelatin. Image taken from [74] with the copyright permission from Elsevier

Pro–Arg–Gly–Glu–4Hyp–Gly–Pro– [23, 29], illustrated in Fig. 1.9. Gelatin is widely used for food industry due to its transparency, clarity, purity, non-toxicity and non-irritation as well as for pharmaceutical and medical applications [23, 73].

The presence of triple helix in the gelatin structure is the main reason for high strength and poor water swelling. Consequently, blending gelatin with hydrophilic polymers like PVA can improve the swelling properties [75, 76]. On the other hand, PVA/gelatin blends possess high ability to form films and hydrogels, which makes such blends a good material candidate for many biomedical applications [76–78]. Furthermore, Pawde et al. [79] also reported that poor electrical conductivity of PVA could be overcome when blended with gelatin. Gao et al. [80] studied the compatibility of PVA/gelatin blends at various blend ratios including 100:0, 80:20, 70:30, 60:40, 50:50, 40:60, 30:70, 20:80 and 0:100 by weight. Their results showed that 80:20 PVA/gelatin blends had the optimum compatibility resulting from a network of hydrogen bonding between components. As a result, higher T_g at 130.3 °C took place as opposed to 120.7 and 121.9 °C for neat PVA and gelatin, respectively, as well as higher tensile strength increased by 22.91% when compared with that of neat gelatin. Blending PVA with gelatin altered the PVA structures by decreasing both chain space and free-volume dipoles. This resulted in the change of blend polarisation behaviour at different frequencies along with increased electrical properties.

1.2.6 Manufacturing Processes

Solution casting, extrusion and melt blending are the most popular manufacturing processes used to prepare neat PVA and PVA blend films, as listed in Table 1.2. In solution casting process, PVA and/or other polymers are dissolved in a suitable solvent like water and acidic solution at specific temperature levels, depending on the properties of polymers, and by continuous mixing to minimise bubbles within

Table 1.2 PVA blends prepared in different manufacturing processes

Polymer blend	Manufacturing process	References
Plasticised PVA	Solution casting	[30, 83, 86]
Plasticised PVA	Extrusion	[87]
Cross-linked PVA	Solution casting	[38, 88]
PVA/ST	Solution casting	[54, 81, 82, 89–98]
PVA/ST	Extrusion	[53, 56, 84, 85, 99–101]
PVA/ST	Melt blending	[52, 102, 103]
PVA/PLA	Extrusion	[8, 104]
PVA/PLA	Melt blending	[62, 63]
Chitosan/PVA/PLA	Solution casting	[59, 105]
PVA/chitin	Solution casting	[66, 67, 106]
PVA/chitosan	Solution casting	[7, 27, 69, 71, 72]
PVA/gelatin	Solution casting	[75, 76, 78, 79, 107, 108]

the solution. Then the completely clear solution is cast in the mould and dried in an oven or under the ambient condition [38, 54, 81, 82]. The same procedure was followed by Mohsin et al. [30, 83] to prepare plasticised PVA films with glycerol and sorbitol by dissolving 5 wt% PVA in 10 ml distilled water at 90 °C via magnetic stirring. Different plasticiser contents were added to completely dissolved PVA by continuous stirring for 6 h in order to produce a homogenous solution, which was then cast on PTFE plates and dried at 80 °C in a vacuum oven. Solution casting has been regarded as the best processing method for manufacturing PVA/ST blends since the 1980s because PVA can easily degrade in melt processing [32]. From an economic viewpoint, solution casting is also an unacceptable manufacturing process as a result of limited efficiency and relatively high cost when compared with thermoplastic processing using extrusion or melt blending [32, 84, 85], especially for mass production.

In an extrusion process, the weighed amounts of polymers with other additives are blended at room temperature (i.e. dry blending) [53] or at an elevated temperature (i.e. melt blending) [8], which are then fed to a single-screw extruder [84, 85, 99] or a twin-screw extruder [53, 56]. A sophisticated design was carried out for extrusion processing parameters such as screw speed and temperature profile from different zones starting from the feeder to the die [53, 84]. The extruded pellets can be co-extruded to a flat die [56], blown [53] and cold pressed [8] to produce films. The screw extrusion is hard to use for processing PVA/ST blends due to their unique rheology in terms of moisture content and processing temperature [84, 85]. On the other hand, PVA is hard to be processed in extrusion because its processing temperature is very close to both degradation and melting temperatures [2]. Consequently, solution casting is considered as the most suitable method for preparing PVA blends.

Melt blending process is used to prepare products with their final form. Polymers are mixed in the dry form and then blended together through a mixer (e.g. Brabender mixer) [62] with a blade-type rotary to melt them at a pre-defined temperature, rotating speed and time [52]. Then molten blends are hot pressed in the mould at the specific temperature, pressure, time and dimensions [63, 102]. Tănase et al. [52] found that there was a relationship between material formulation of plasticised PVA/ST blends and melt processing parameters. Their results showed that increasing the ST content from 0 to 30 wt% in the blends increased the melt viscosity when compared with plasticised PVA blends leading to an increase in power consumption with much lower processability. Overall, PVA/ST blends at a low ST content are much easier to be processed in melt blending process.

1.2.7 Properties

As mentioned earlier, most PVA properties depend on molecular weight and HD [3, 32], which can be modified by using blending processes. Polymer blending is an effective method to produce a material system with better properties, as opposed to their individual components, in order to avoid inherent limitations of neat polymers [109]. Therefore, blending PVA with other polymers, particularly those having similar solubility parameters, can improve overall blend material performance owing to the formation of strong hydrogen bonds instead of weak van der Waals interaction [49].

Material morphological structures should be analysed carefully as a key indicator for many other material properties. Neat PVA films possess smooth and homogenous surfaces with some irregularities along the cross section, reflecting typical semicrystalline structures of PVA [110, 111], as depicted in Fig. 1.10.

Smooth and homogenous morphological structures disappear when PVA is blended with ST particularly when the PVA content is equal to or higher than the ST content due to their partial miscibility consisting of ST-rich phase distributed in PVA-rich phase [112], Fig. 1.11a. This structure is completely changed when plasticising PVA/ST blends with glycerol due to the enhanced compatibility between polymers [112]. Moreover, an excessive amount of plasticisers may form oily layers on the film surface, as evidenced by a blooming/blushing phenomenon under SEM examination [89], Fig. 1.11b. The presence of GL helps to improve the compatibility between PVA and chitosan as well to produce smooth and homogenous blend surfaces [105, 113]. As such, the presence of plasticisers is beneficial to increase the compatibility. However, a higher plasticiser content can also cause phase separation while the lower content induces the hardening effect instead of the plasticisation.

Neat full-hydrolysis PVA has maximum quantities of tensile strength, Young's modulus and elongation at break about 1.6 GPa, 48 GPa and 6.5%, respectively [4], as opposed to 25.4 MPa, 27.6 MPa and 260% accordingly for partial-hydrolysis PVA [114]. Blending PVA with PLA can improve the tensile strength and water resistance compared with neat PVA counterparts [62, 104]. Li et al. [8] reported that the tensile strength of plasticised PVA/PLA blends was increased by 11.86% as

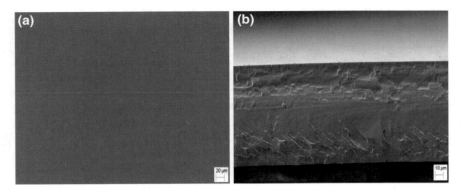

Fig. 1.10 SEM images of neat PVA: **a** surface structure and **b** cross-sectional structure. Image taken from [110] with the copyright permission from Elsevier

Fig. 1.11 SEM images for **a** un-plasticised PVA/ST blends (image taken from [112] with the copyright permission from Springer) and **b** over-plasticised PVA/ST blends (image taken from [89] with the copyright permission from Tayler and Francis)

compared with that of plasticised PVA counterparts, while the elongation at break was decreased by 14.81%. Furthermore, the water contact angle of blends increased slightly as a sign of the improvement of water resistance. A similar behaviour of improving tensile strength and reducing elongation at break was also observed when blending PVA with chitosan due to the interactions between hydroxyl groups of PVA with NH_2 and hydroxyl groups of chitosan [70]. On the other hand, blending PVA with ST improved Young's modulus and reduced the tensile strength and elongation at break owing to high brittleness and amorphous nature of ST regardless of its sources [115]. This behaviour appeared to be more pronounced with increasing the ST content in the blends [95, 116]. Consequently, many studies have focused on using cross-linking agents like sodium benzoate, borax, citric acid, boric acid, glutaraldehyde and tetraethylene glycol diacrylate to improve mechanical properties of PVA/ST blends [82, 91, 93]. These agents form strong intermolecular linkages between the components by reacting with the hydroxyl groups of polymers leading to the improvement of tensile strength [82, 97]. Besides, plasticisers like urea and different polyol-based materials are used to enhance the flexibility and elongation at break for blends by penetrating between polymeric chains in order to improve their mobility and free volumes [89, 90]. In other words, mechanical properties of blends depend on the properties of individual components, as well as the interactions between them.

Thermal properties of neat PVA and its blends can be determined with the aid of thermogravimetric analysis (TGA) and DSC. The T_m of neat PVA is determined to be 230 °C, while the T_g becomes a function of HD. A full-hydrolysis PVA has a T_g around 85 °C as opposed to 58 °C for partial-hydrolysis PVA [117]. PVA can thermally degrade in both molten and solid states. The high flexibility and chain mobility of PVA in a molten state promote the fragmentation and elimination of chain segments, which produces saturated and unsaturated volatile ketones and aldehydes, as well as water [117]. In the solid state, the side groups are eliminated and followed by the reduction of melting temperature and degree of crystallinity with the production of appreciable amounts of polyenes in isolated and conjugated forms, as well as small amounts of carboxyl groups [117]. Neat PVA films and PVA blends prepared by solution casting possess similar thermal degradation steps. The initial step includes 10% weight loss due to the loss of bonded water at around 100 °C. The second step lies in 70% weight loss as a result of the degradation process with dehydration, polymer scission and decomposition in a temperature range of 150–380 °C. Further, the third step comprises the formation of end products taking place at a temperature range of 380–500 °C [112]. In comparison with PVA/PLA blends and PVA/chitin blends (see Sects. 2.2 and 2.3), PVA/ST blends do not show a single and clear T_g because of the partial miscibility of polymers in the absence of plasticisers [111]. Furthermore, the presence of plasticisers decreases the T_g and T_m, as well as degree of crystallinity and thermal enthalpies of both neat PVA and PVA blends [90]. Generally, these plasticiser molecules are smaller than polymer molecules that make them easier to penetrate polymeric chains and reduce cohesion forces between polymer molecules leading to the increase in the mobility of polymeric chains [95].

Neat PVA is not a completely biodegradable polymer in all environments, particularly in the absence of specific conditions such as relative humidity, temperature, pH level and types of microorganisms [87]. Neat PVA is completely biodegradable in activated sludge, while its general degradation rate is in the range of 8–9% over the period of 74 days in soil, which was increased up to 13% in an aerobic condition on the 21st day [2, 87]. Consequently, neat PVA is blended with other polymers like ST and chitin to improve the biodegradability in different environments. Lots of studies have showed that ST can be attacked and consumed first by the microorganisms in PVA/ST blends in soil burial biodegradation tests because it is a completely biodegradable polymer. Since porous residues of PVA films are left behind, weak structures can be fragmented easily [21, 25, 118, 119]. A similar behaviour can also be identified in PVA/chitin blends. Takasu et al. [67]

reported that the residual weight of soil burial films was decreased after 150 days from 89% for neat PVA films to 24% for PVA/chitin films. The biodegradation of PVA blend films depends on the blend components, miscibility of these components and types of biodegradation media [61].

Neat PVA as a water-soluble polymer has poor water resistance in terms of water solubility (W_s), water uptake (W_a), and water contact angle, as well as WVP [2]. However, it has good gas barrier properties due to dense, small and closely packed crystallites [33]. Moreover, barrier properties of PVA blends can be affected by temperature, relative humidity, chemical contents such as plasticisers and cross-linking agents, as well as hydrophilicity of blend components [120]. For instance, WVP and W_s of most PVA/ST films were increased linearly with increasing the ST content in the blends because of high hydrophilicity of ST, particularly in the presence of plasticisers [89, 111, 116]. Furthermore, PVA/chitin blends show higher hydrophilicity when compared with neat PVA, as evidenced by decreasing water contact angle [67]. According to Liu et al. [72], blending PVA with chitosan reduced the gas permeability due to strong interactions between the components of a strong packed structure, while the WVP of blends was increased as the hydrophilicity of blends was enhanced to improve the diffusivity of water molecules within the films. As such, barrier properties of PVA are not significantly improved when blended with other biopolymers, which, however, can be altered with the incorporation of hydrophobic fillers.

1.2.8 Applications

The applications of neat PVA and PVA blends have been significantly increased during last three decades according to Scopus Database shown in Fig. 1.12.

Most of these applications are related to medical fields due to their biocompatibility, non-toxicity and easy formability particularly in wound healing, drug delivery systems, contact lenses and artificial organs [121, 122]. For example, Liu et al. [60] prepared PVA/PLA membranes with high miscibility between components leading to the improvement of Young's modulus by 170% as opposed to that of neat PVA, which have been successfully used as high-performance tissue scaffolds in biomedical areas. Moreover, Li et al. [8] found the tensile strength of PVA/PLA blends increased slightly by 11.86% as compared with neat PVA, as well as slight reduction in elongation at break by 12.9%. On the other hand, the water contact angle of blends improved by three folds relative to neat PVA counterpart leading to possible applications for food packaging materials.

Rafique et al. [123] proved that blending PVA with chitosan could greatly improve chemical and antimicrobial properties of their blends, as well as reduce the production cost for manufacturing more competitive materials in biomedical sectors. This was validated experimentally by Liu et al. [124] when preparing PVA/chitosan hydrogel for wound healing with excellent antimicrobial activities against *E. coli*. Similar results were also reported by Zhao et al. [125] resulting in

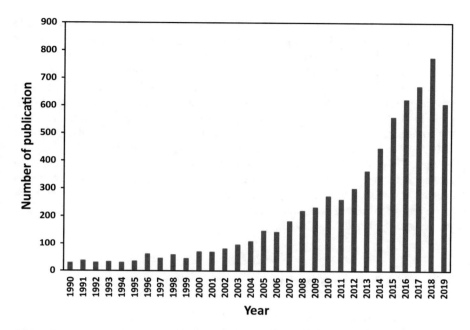

Fig. 1.12 Number of publications related to the applications of neat PVA and PVA blends in medical sectors

homogenous morphological structures of PVA/carboxymethylated chitosan (CM-chitosan) hydrogels, which were free of phase separation and/or precipitation up to 8 wt% CM-chitosan, as well as demonstrated a linearly increasing trend in tensile strength of blends, as compared with that of neat polymers. Costa-Júnior et al. [126] found similar homogenous morphological structures of PVA/chitosan hydrogels with a low swelling rate by 150% as opposed to that of neat PVA and high cell viability of 97%, which could be successfully applied in skin engineering field. Zhuang et al. [127] prepared different PVA/chitosan membranes with weight blend ratios of 3:1, 2:1, 1:1, 1:2 and 1:3. Their findings revealed that tensile strength and flexibility improved linearly with increasing the PVA content, while cytotoxicity and microbial growth decreased with increasing the chitosan content leading to their potential applications in guided tissue regeneration (GTR). On the other hand, Bonilla et al. [128] reported that blending PVA with 20 wt% chitosan enabled to significantly improve tensile strength and Young's modulus by 86.95 and 428.51%, respectively, because of enhanced interactions between them. Furthermore, homogenous morphological structures of PVA/chitosan blends improved the barrier properties against light and ultraviolet radiation with remarkable inhabitation of microbial growth for the direct benefit to be used as food packaging materials.

Gao et al. [129] studied morphological structures of PVA/gelatin blends by Fourier transform infrared (FTIR) spectroscopy, X-ray diffraction (XRD) and SEM. Their results demonstrated that PVA/gelatin blends with intense interactions and compatibility could be used for a wide range of biomedical applications. For example, Hago and Li [130] reported that tensile strength and Young's modulus of PVA/gelatin blends were increased by 60.97 and 77.78%, respectively, as opposed to those of neat PVA with the addition of 2 wt% gelatin. These results were ascribed to dense and rigid morphological structures of blends, as well as good swelling properties for the healing dressing of exudative wounds. Moreover, Wang et al. [131] found hydroxyapatite/PVA/gelatin blends had high porosity and water absorption rate by 78 and 312.7%, respectively, as compared with those of neat polymers with non-cytotoxicity behaviour and high compatibility in vivo and in vitro. They could be used as water-swellable hydrogels for scaffolds in cartilage tissue engineering. Furthermore, Fan et al. [132] prepared PVA/chitosan/gelatin blend hydrogels for wound dressing with higher swelling properties by 20–40 folds as opposed to those of chitosan/gelatin counterparts, which arose from hydrophilic properties of PVA to stop bleeding.

1.3   PVA Nanocomposites

Polymer nanocomposites consist of nanoscaled reinforcement constituents uniformly dispersed in continuous polymer matrices (i.e. neat polymers or polymer blends) [133]. Such nanofillers can be categorised as one-dimensional (nanoplatelets), two-dimensional (nanotubes) and three dimensional (spherical nanofillers) [133]. Superior improvements in material properties such as barrier, mechanical and thermal properties of polymer nanocomposites can be achieved at small nanofiller contents (≤ 5 wt%) when compared with conventional polymer composites reinforced with 40–50 wt% conventional fillers [13, 29]. On the other hand, polymer nanocomposites still encounter the limitations in relation to the dispersion of nanofillers during the processing, as well as nanofiller agglomeration at the high contents, as shown in Fig. 1.13 [134]. Biodegradable polymers are used widely as the matrices of polymer nanocomposites to enhance their relatively low mechanical and thermal properties, as well as poor barrier properties in contrast to petroleum-based polymers [135]. In particular, PVA is regarded as good matrices for polymer nanocomposites when used in medical and material packaging applications.

1.3.1   PVA/Montmorillonite (MMT) Nanocomposites

MMTs are the most popular nanoclays in the family of layered silicates used as nanofillers in polymer nanocomposites [133]. MMTs have a 2:1 layered structure consisting of one crystal sheet of alumina octahedron sandwiched between two crystal sheets of silica tetrahedron [1, 10], Fig. 1.14. These layers have the ability to organise themselves in stacks with regular van der Waals gaps, which are known as

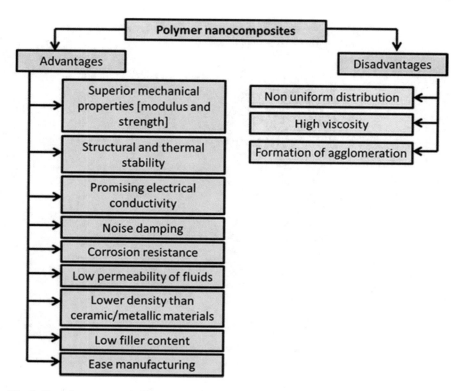

Fig. 1.13 Advantages and disadvantages of polymer nanocomposites. Image taken from [134] with the copyright permission from Springer

galleries or interlayers with the thickness of approximately 1 nm and the length in range of 30 nm to several microns. These van der Waals forces are relatively weak, so small molecules like water, chemicals and monomers, as well as polymers could easily penetrate between them and expand MMT interlayers [1, 29, 136]. MMTs are inorganic materials with hydrophilic nature due to the presence of hydrated sodium and potassium ions to improve the miscibility with hydrophilic polymers such as PVA [10, 29]. The unique characteristics of MMTs like a high surface area of 750 m^2/g and high aspect ratios in range of 50–1000 can greatly enhance mechanical and thermal properties of polymer/MMT nanocomposites [13].

The resulting properties and structures of polymer/clay nanocomposites depend on nanoclay dispersibility within polymer matrices [1]. Three different structures of polymer/clay nanocomposites can be produced based on material selection and preparation methods, as shown in Fig. 1.15 [10, 136]:

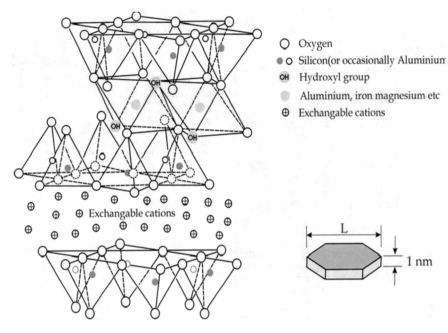

O   Oxygen

● O   Silicon(or occasionally Aluminium

OH   Hydroxyl group

Aluminium, iron magnesium etc

⊕   Exchangable cations

Fig. 1.14 Chemical structure and size of MMTs. Image taken from [1]

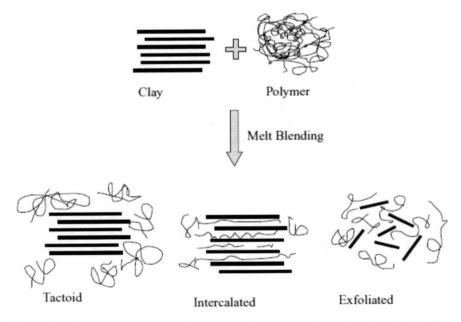

Clay   Polymer

Melt Blending

Tactoid   Intercalated   Exfoliated

Fig. 1.15 Polymer/clay nanocomposite structures. Image taken from [16] with the copyright permission from Wiley

- **Tactoid structure**: It is also known as non-intercalated structure [136], non-mixing composites or microcomposites [1]. In this structure, polymeric molecules cannot diffuse between clay interlayers due to no interaction between them, generally resulting in conventional microcomposite structures [16].
- **Intercalated structure**: Polymeric chains are intercalated between clay interlayers and expanded with d-spacing values of 2–5 nm to produce multilayered morphological structures comprising alternative polymeric chains and clay layers [1, 10].
- **Exfoliated structure**: Clay interlayers are expanded by more than 5–10 nm when individual and randomly oriented clays are homogeneously dispersed within continuous polymer matrices to yield completely exfoliated structures [1, 10].

Both intercalated and exfoliated structures can be found in PVA/MMT nanocomposites [1]. Furthermore, XRD results showed that the d-spacing of MMTs within PVA/MMT nanocomposites was increased up to 1.92 nm when compared with that of pristine MMTs, and the degree of intercalated structures was increased with increasing the MMT content. Moreover, associated results determined by transmission electron microscopy (TEM) demonstrated clear exfoliated structures at the MMT contents of 0.1 and 0.2 wt% though the degree of exfoliated structures was decreased with increasing the MMT content [137]. Furthermore, Majdzadeh-Ardakani and Nazari [138] reported that the tensile strength of PVA/ST/MMT nanocomposites was increased by 30.18% with the incorporation of 4 wt% MMTs when compared with that of neat polymers due to the formation of exfoliated/intercalated structures. Nonetheless, the tensile strength of such nanocomposites declined beyond 4 wt% MMTs in the absence of exfoliated clay structures, particularly at high MMT content levels when typical MMT agglomeration took place. In short, the resulting structures of polymer/clay nanocomposites depend on nanofiller dispersion within polymer matrices. In other words, well-dispersed nanofillers can produce exfoliated structures particularly at low nanofiller contents, while such exfoliated structures can be replaced by intercalated counterparts with increasing the nanofiller content.

1.3.2 PVA/Halloysite Nanotube (HNT) Nanocomposites

HNTs are considered as the other popular nanoclays that were discovered first by Omalius d'Halloy in 1826 [31, 139]. HNTs are naturally available nanoclays formed by the hydrothermal alteration of alumina silicate minerals. HNTs can exist in different forms depending on their formations [139]. The tubular form of HNTs is more popular than spherical and plate-like counterparts with typical inner diameters of 5-20 nm, outer diameters of 10–50 nm and lengths of 500 nm–1.2 μm depending on their sources [140, 141]. HNTs have a chemical formula $Al_2Si_2O_5(OH)_4 \cdot nH_2O$ similar to kaolin with a monolayer of water [140, 141]. The

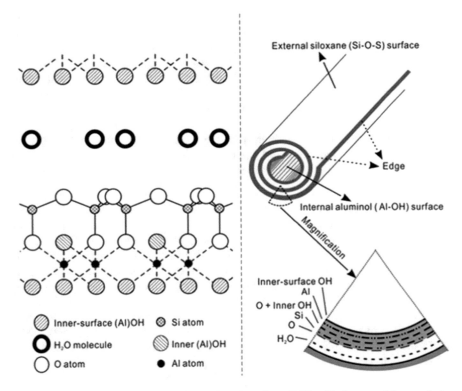

Fig. 1.16 Chemical structure of HNTs. Image taken from [143] with the copyright permission from Elsevier

material classification of HNTs depends on n values with $n = 2$ and d-spacing of 10 Å for hydrated HNTs, as well as $n = 0$ and d-spacing of 7 Å for dehydrated HNTs [142, 143]. HNTs have a unique 1:1 crystalline structure consisting of tetrahedral sheets with corner-shared SiO_4 and octahedral sheets with edge-shared AlO_6 in possession of low hydroxyl groups on the outer surfaces to improve their dispersion and reduce tube-to-tube interactions [4, 141], as illustrated in Fig. 1.16. There are several scenarios to increase the dispersibility of HNTs particularly at high contents when HNTs tend to agglomerate because of their high surface areas and reactivity, though HNTs have good dispersibility compared with other nanotube fillers to get better interfaces in polymer/HNT nanocomposites [144]. Most of these scenarios are based on chemical surface modification for HNTs by using coupling agents, surfactants, intercalated with salts and organic compounds, as well as warping with polymers or inorganic layers [144].

As compared with other nanofillers with similar tubular structures like carbon nanotubes (CNTs), boron nitride nanotubes and metal oxide nanotubes, HNTs are considered as relatively cheap nanofillers with abundant availability, and can be dispersed easily within polymer matrices [142, 143, 145]. Furthermore, HNTs are a good nanofiller candidate for many nanocomposite systems due to their high aspect ratios, low hydroxyl density on the surfaces, as well as good mechanical properties and thermal stability [143, 145]. HNTs have been identified to show their non-cytotoxicity with high viable rate of cells in vivo and in vitro over decades [144, 146, 147]. In addition to high biocompatibility, non-carcinogenicity and non-toxicity, PVA/HNT nanocomposites can also be widely utilised in biomedical applications such as artificial heart surgery, drug delivery systems, contact lenses, scaffolds for tissue engineering, tumour cell isolation and wound dressing [4, 146]. In particular, Khoo et al. [140] found the addition of 0.5 wt% HNTs improved the tensile strength and tear strength, as well as the swelling resistance of PVA/chitosan/HNT nanocomposites as favourable packaging materials. Moreover, PVA/HNT nanocomposites can be used as flame retardant materials due to tubular structures of HNTs to trap flammable volatile molecules [148]. Furthermore, well-dispersed HNTs within polymer matrices can play an insulating role on composite surfaces and get rid of flammable gases from the surfaces. In particular, HNTs have high thermal stability reaching 85.5 wt% residues at 800 °C, resulting in the improvements of both thermal stability and flame retardancy of nanocomposites [144]. Based on intrinsic properties and widespread medical applications, HNTs were chose in this work as nanofillers of bionanocomposite films, especially targeting food packaging applications.

1.3.3   PVA/Graphene Oxide (GO) Nanocomposites

Graphene oxides (GOs) were discovered for the first time by a British chemist B. C. Brodie during the chemical treatment of graphite in 1859. The original name of GOs was graphite oxides, but it was changed to graphene oxides in 2004 [149, 150]. Notwithstanding that there are many preparation methods for the bulk production of GOs, modified Hummer's method is still considered as the most popular one with the higher oxidation degree [150, 151]. GOs in a chemical formula of $C_{11}H_4O_5$ are two-dimensional nanosheets with the thickness of 1 nm and lateral dimensions in range of a few nanometres to several microns with high aspect ratios [149]. GOs have various oxygen functional groups such as epoxide, hydroxyl, carboxyl and carbonyl groups on the edges and basal planes [9, 150, 152], as shown in Fig. 1.17. These oxygen groups provide strong interactions to GOs with other polar-molecule-based polymers to yield intercalated and/or exfoliated nanocomposite structures [152, 153]. Moreover, polymer/GO nanocomposites have much higher improvements in mechanical and barrier properties, electrical and thermal conductivities and acceptable transparency [153]. On the other hand, these oxygen groups can be reduced thermally (by thermal annealing or thermal irradiation) or chemically (by reducing reagents such as hydrate and hydrazine, photocatalyst reduction or solvothermal reduction) to produce reduced GOs (RGOs) with better properties as compared with pristine GOs [154, 155]. RGO sheets have other

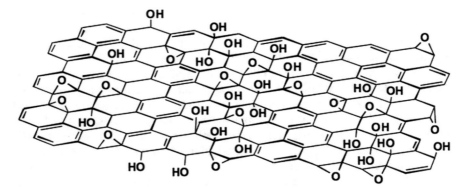

Fig. 1.17 Chemical structure of GOs. Image taken from [153] with the copyright permission from Wiley

alternative names such as chemically modified graphene, functionalised graphene and chemically converted graphene [155]. Aslam et al. [156] studied the effect of GOs and RGOs on mechanical and thermal properties of PVA/0.5 wt% GO nanocomposites and PVA/0.5 wt% RGO nanocomposites, respectively. Their results showed that tensile strength and X_c of PVA/RGO nanocomposites were increased by 80 and 25.5%, respectively, as compared with 68 and 14.6% for PVA/GO nanocomposites. Furthermore, T_m, T_g and T_c of PVA/RGO nanocomposites were improved by 27.9, 9.8 and 42.7 °C, respectively, in accord with 9.5, 6.5 and 29.2 °C for PVA/GO nanocomposites.

Huang et al. [9] found that the oxygen permeability (OP) and WVP of PVA/GO nanocomposite films were reduced by 98.86 and 67.91%, respectively, with increasing the GO content from 0 to 1 wt%. This phenomenon was associated with full GO exfoliation in polymer/GO nanocomposites to generate more tortuous paths. Accordingly, PVA/GO nanocomposites were proven to be used as effective packaging materials with good barrier properties [151]. Moreover, Sellam et al. [157] and Liu et al. [158] reported remarkable improvements in tensile strength and Young's modulus of PVA/GO nanocomposites relative to those of neat PVA due to inherently high Young's modulus of GOs, as well as strong interactions between functional groups of GOs with hydroxyl groups in polar PVA.

1.3.4 PVA/Carbon Nanotube (CNT) Nanocomposites

Since the first discovery of CNTs in 1991, they become very popular one-dimensional carbon-based nanofillers in possession of high aspect ratios over 1000, excellent mechanical properties and high thermal and electrical conductivities [159–161]. CNTs are built up by using sp^2 carbon–carbon bonds in a layered structure with strong in-plane bonds and weak out-of-plane van der Waals interactions [31]. CNTs are available in three major forms, namely single-walled CNTs (SWCNTs), double-walled CNTs (DWCNTs) and multi-walled CNTs (MWCNTs), as depicted in Fig. 1.18. SWCNTs consist of a single graphite crystal that can be rolled into a cylinder form, while DWCNTs have two graphite crystals rolled concentrically. On the other hand, MWCNTs comprise concentric multi-graphite crystals that are rolled over around central hollow structures with an interlayer spacing of 0.34 nm [159, 161, 162]. CNTs are used widely to produce polymer/CNT nanocomposites in many applications, particularly for electromagnetic interface shielding with the combination of high electrical conductivity of CNTs and good flexibility of polymers [159, 163]. Polymer/CNT nanocomposites possess good mechanical, thermal and electrical properties due to inherently high mechanical properties (50–100 GPa in tensile strength and 50–1000 GPa in Young's modulus), high thermal conductivity (2×10^3 W/mK for SWCNTs and 3×10^3 W/mK for MWCNTs), high electrical conductivity (10^4 S/cm for SWCNTs and 1.85×10^3 S/cm for MWCNTs), as well as good fire retardancy and high barrier properties of CNTs [160, 163]. There are many challenges encountered by using polymer/CNT nanocomposites due to high surface stability of CNTs resulting in the prevention of interactions between CNTs and polymer matrices. Moreover, small sizes and high surface areas of CNTs increase the possibility of CNT agglomeration, as compared with well-dispersed CNTs within polymer matrices. Chemical modifications of CNTs with the incorporation of functional groups can overcome some of these challenges by reducing van der Waals interactions [159, 160, 163]. For example, Paiva et al. [164] used N,N'-dicyclohexyl carbodiimide-activated esterification reaction to functionalise SWCNTs with PVA in order to improve their dispersion within polymer matrices. Their results of mechanical properties showed that yield strength and Young's modulus of PVA/ functional SWCNT nanocomposite films increased linearly by 54.21 and 55.0%, respectively, with increasing functional SWCNT contents from 0 to 5 wt% as opposed to those of PVA/pristine SWCNT composite films. Similarly, Lin et al. [165] studied the morphological structures of PVA/functional CNT nanocomposites after using the same esterification reaction to improve the dispersion of SWCNTs and MWCNTs within polymer matrices. Homogenous structures without phase separation occurred in possession of uniform nanofiller distribution according to TEM images depicted in Fig. 1.19.

Similarly, Liu et al. [166] functionalised SWCNTs with multi-surface hydroxyl groups to improve their dispersion within PVA matrices. Single T_g of PVA/ functional SWCNT nanocomposites was indicative of improving the interactions between polymer matrices and SWCNTs. Moreover, yield strength and Young's modulus, as well as T_g of PVA/functional SWCNT nanocomposites were found to be increased by 47%, 79% and 5.5 °C, respectively, when compared with those of neat PVA. Basiuk et al. [167] reported that SWCNTs could be more easily dispersed within cross-linked PVA matrices in comparison with MWCNTs with relatively high van der Waals interactions. Consequently, Wongon et al. [168] used sodium dodecylbenzenesulfonate (NaDDBS) dispersant agent to reduce the surface tension between MWCNTs and improve their dispersion within PVA solution. First

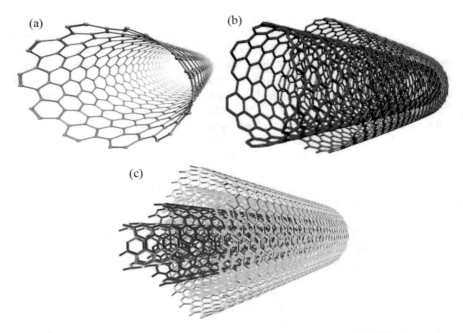

Fig. 1.18 CNT structure configurations: **a** SWCNT, **b** DWCNT and **c** MWCNT. Image taken from [160] with the copyright permission from Tayler and Francis

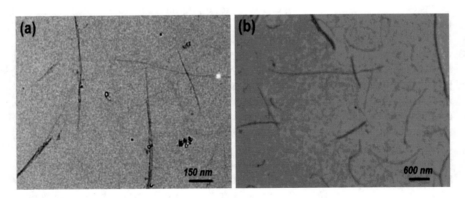

Fig. 1.19 TEM images of **a** PVA/functional SWCNT nanocomposites and **b** PVA/functional MWCNT nanocomposites. Image taken from [165] with the copyright permission from American Chemical Society

of all, NaDDBS was dissolved in deionised water using a magnetic stirrer for 20 min. Then weighted amounts of MWCNTs were added to the solution via ultrasonication for 5 min. NaDDBS consists of hydrophobic and hydrophilic components. The hydrophilic component was dissolved in water, while hydrophobic component was adsorbed on MWCNTs to reduce their surface tension and aggregation leading to better dispersibility within PVA solution. In addition, Ryan et al. [169] found that Young's modulus of PVA/1 wt% MWCNT nanocomposites was increased by 5.8 folds as opposed to that of neat PVA. However, when increasing the MWCNT content beyond 1 wt%, Young's modulus of such nanocomposites was decreased instead, which was ascribed to typical MWCNT agglomeration issue.

1.3.5 PVA/Cellulose Nanocomposites

Cellulose is a linear chain polymer in polysaccharide family with abundant resources on earth [31, 170]. Cellulose with a chemical formula $(C_6H_{11}O_5)_n$ consists of ringed glucose molecules connected by $\beta(1\rightarrow4)$ linkages, as shown in Fig. 1.20. The interactions between hydroxyl groups and oxygen of adjacent ring molecules via hydrogen bonding improve cellulose stability and build its linear chains with the diameter of 2–20 nm and the length of a few microns [170–172]. Consequently, cellulose is regarded as an insoluble polymer in water owing to its high density of hydrogen bonding networks. Soluble cellulose in water can be synthesised by the chemical functionalisation through the esterification or etherification process with full or partial replacement of hydroxyl groups in cellulose with ester or ether groups, respectively [171].

Cellulose can be extracted from a wide range of natural resources such as bacteria, algae, tunicate and wood, as well as plants like cotton, sisal, wheat straw, flax, potato tubers, etc. [170]. Depending on the biosynthesis process used to extract cellulose from these resources, a range of cellulose forms can be achieved like cellulose nanocrystals (CNCs), nanofibril celluloses (NFCs), cellulose nanowhiskers (CNWs), cellulose nanoparticles (CNPs) and cellulose microfibrils (CMFs) [173, 174]. All these types of cellulose fillers are biodegradable, biocompatible and cheap with abundant availability, low densities, high aspect ratios and large surface areas. Consequently, they can be deemed as effective nanoreinforcements in polymer nanocomposite systems [173]. Cai et al. [175] discovered that there was an intimate interfacial interaction between NFCs and PVA associated with hydrogen bonding between hydroxyl groups for improving mechanical, thermal and optical properties. For instance, Spagnol et al. [176] used CNWs to improve the mechanical properties of PVA/CNW nanocomposites. Their results showed that storage modulus, Young's modulus and tensile strength of PVA/CNW nanocomposites were increased linearly by 60.28, 114.92 and 16.62%, respectively, with increasing the CNW content from 0 to 9 wt% due to strong hydrogen and interfacial bonding for better load transfer. Popescu [177] identified similar strong interactions between PVA and CNC via hydrogen bonding to improve the crystallinity degree of PVA/CNC nanocomposites by 127.68% when compared with neat PVA leading to decreasing the water absorption rate of nanocomposites by 19.35% because of the consumption of free hydroxyl groups. Cheikh et al. [178]

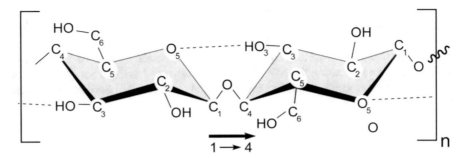

Fig. 1.20 Chemical structure of cellulose. Image taken from [170] with the copyright permission from Royal Society of Chemistry

reported that good compatibility between PVA and NFCs yielded exfoliated structures in nanocomposites. As a result, Young's modulus and tensile strength of PVA/10 wt% NFC nanocomposites were increased by 113 and 41%, respectively, as compared with those of neat PVA counterparts. Moreover, the decomposition temperature and T_g of PVA/10 wt% NFC nanocomposites were increased by 66.5 and 3.6 °C, respectively when benchmarked against those of neat PVA. Such results were believed to be caused by the restriction of chain mobility of PVA in the presence of NFCs. Similar findings were reported by Voronova et al. [179] with increasing decomposition temperatures of PVA/CNC nanocomposites up to 80°C, as compared with those of neat PVA and CNCs at loading range from 8 to 12 wt%. On the other hand, these improvements in thermal stability declined when incorporated with 12 wt% CNCs. Ibrahim et al. [174] found that the incorporation of CNPs into PVA matrices increased the biodegradation rate of nanocomposite films thanks to the complete biodegradable nature of CNPs.

1.3.6   Manufacturing Processes

Depending on nanocomposite constituents and processing steps, manufacturing methods of nanocomposites can be summarised as follows:

- **In situ polymerisation**: It is also called interlamellar polymerisation. In this process, nanofillers (mostly nanoclays) are combined with the monomer solution or liquid monomers to swell prior to the polymerisation to produce exfoliated nanocomposites, Fig. 1.21a. Radiation, heat and catalysts can be used as polymerisation initiators. This method can be restricted by the unavailability of suitable monomers [10, 29, 136, 180].
- **Melt intercalation**: This method is known as direct melt intercalation as well. Nanofillers are melt compounded with polymer matrices (mostly thermoplastic polymers) above the softening point of polymers with the aid of extruders or

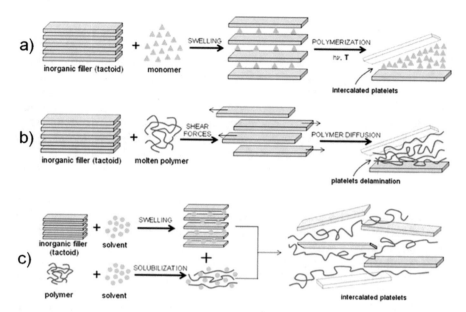

Fig. 1.21 Typical preparation methods of polymer nanocomposites: **a** in situ polymerisation, **b** melt intercalation and **c** solution intercalation. Image taken from [180] with the copyright permission from Royal Society of Chemistry

other internal mixers. Polymeric molecules penetrate between nanofillers to produce intercalated and/or exfoliated nanocomposites, as demonstrated in Fig. 1.21b [10, 29, 136, 180]. This method is considered as an ecofriendly process in the absence of solvents [133]. Although it becomes a standard manufacturing method at an industrial upscaling level for polymer/clay nanocomposites [29], unfortunately it is still restricted to thermoplastics [180].

- **Solution intercalation**: It is also known as solution processing or solution casting. The nanofillers are dispersed in water or other suitable chemical solvents with the aid of mechanical and magnetic mixing, as well as sonication mixing. In the meantime, the polymer is dissolved in water or other solvents. Then the suspension of nanofillers is mixed with polymer solution by way of continuous agitation, which is followed by casting in flat surfaces and drying to evaporate the solvents, Fig. 1.21c [10, 29, 133, 180]. This method is suitable for most nanofiller types such as nanoclays particularly HNTs, as well as CNTs and GOs [31]. In general, it is used to produce nanocomposites in thin films with predominantly intercalated structures on a laboratory scale. However, from an industrial point of view, solution intercalation is not ecofriendly and economically prohibitive with the existence of chemical solvents [136].

1.3.7 Properties

In view of packaging applications, the improvements of mechanical, thermal and barrier properties of polymer nanocomposites are of a major concern, which highly depend on nanofiller content, nanofiller size, aspect ratio and nanofiller dispersibility within polymer matrices, as well as filler–matrix interfacial bonding [181, 182]. Chemical modification of nanofillers is one of the most interesting ways to improve their dispersibility within polymer matrices, thus leading to remarkable improvements in nanocomposite properties. For example, Spagnol et al. [176] modified CNWs with maleic anhydride (MA), succinic anhydride (SA), acetic anhydride (AA) and phthalic anhydride (PA). Subsequently, modified fillers were employed to compare the effect of modification additives on mechanical properties of PVA/modified-CNW nanocomposites relative to those of neat PVA. Their results revealed that pristine and modified CNWs with different additives highly improved tensile strength and Young's modulus of nanocomposites relative to neat PVA due to strong interfacial bonding between polymer matrices and fillers despite an evident decrease in elongation at break when excluding PVA/CNW-MA nanocomposites and PVA/CNW-AA nanocomposites. Such a finding was attributed to relatively small particle sizes of these additives to promote the mobility of polymeric chains (see Table 1.3). Moreover, Asad et al. [183] found well-dispersed oxidised nanocellulose helped to drastically improve tensile strength and Young's modulus of PVA/oxidised nanocellulose composites by 122 and 291%, respectively, with increasing the nanofiller content from 0.5 to 4 wt%. Strong networks of hydrogen bonding appeared to make a great contribution between components along with the reduction of elongation at break by 42%. On the other hand, mechanical properties of PVA/oxidised nanocellulose composites declined slightly beyond the oxidised nanocellulose content of 4 wt% due to the agglomeration of nanofillers. Loryuenyong et al. [114] found that the tensile strength and Young's modulus of PVA/GO nanocomposites were increased by 18.11 and 76.44%, respectively, when incorporated with 1.5 wt% GOs owing to the inherently high strength of GOs. Nevertheless, such properties tended to diminish with increasing the GO content beyond 1.5 wt% resulting from typical nanofiller agglomeration. Apparently, morphological structures of nanocomposites play an important role in their final resulting material properties. Strawhecker and Manias [181] demonstrated that exfoliated structures of PVA/MMT nanocomposites yielded the significant improvement of Young's modulus by 300% with the incorporation of 4 wt % MMTs when compared with that of neat PVA. Increasing the MMT content from 4 to 10 wt% induced the decrease in Young's modulus of such nanocomposites with the highly stacked and aggregated MMTs. Similarly, Koosha et al. [184] found exfoliated structures were the main reason behind improving mechanical properties of PVA/chitosan/Na-MMT nanocomposites, as compared with those of polymer matrices. Their results showed that tensile strength increased by 193.72 and 187.07% with the addition of 1 and 3 wt% Na-MMTs, respectively, when compared with that of PVA/chitosan blends, while Young's modulus was improved by

Table 1.3 Mechanical properties of PVA/CNW nanocomposites [176]

Nanocomposite sample	Young's modulus (kPa)	Tensile strength (kPa)	Elongation at break (%)
Neat PVA	90.03	52.94	90.16
PVA/9 wt% CNW	193.5	61.74	49.97
PVA/9 wt% CNW-MA	215.71	69.14	127.62
PVA/9 wt% CNW-SA	174.8	66.59	42.97
PVA/9 wt% CNW-AA	167.34	64.53	135.8
PVA/9 wt% CNW-PA	105.41	63.80	52.62

351.57 and 241.05% accordingly. Feiz and Navarchian [185] reported similar results related to PVA/MMT modified with chitosan (CsMMT) composite hydrogels. Moreover, Khoo et al. [140] suggested that the agglomeration of HNTs could act as stress concentration sites leading to the reduction in tensile strength and Young's modulus, as well as elongation at break for PVA/chitosan/HNT nanocomposite films beyond 0.5 wt% HNTs. The crystallinity level of nanofillers and their good compatibility with polymer matrices could alleviate the effect of agglomeration issues to a great extent. Liu et al. [186] reported that tensile strength and Young's modulus of PVA/NFC nanocomposites were increased in a linear manner by 87.2 and 522.97%, respectively, when increasing the NFC content from 0 to 60 wt% due to the high crystallinity level of NFCs and good compatibility with PVA matrices to yield overall rigid nanocomposite structures in high load-bearing capacity.

The presence of nanofillers can also enhance thermal stability of polymer nanocomposites due to insulating effect and the role of nanofillers as a barrier material against mass transfer in a decomposition process [29]. Based on previous work mentioned by Nistor and Vasile [187], exfoliated PVA/ST/modified-MMT nanocomposites led to increasing the decomposition temperature and char residues as compared with intercalated MMT nanocomposites, which could be related to good dispersion of modified-MMTs to hinder heat and mass transfer of nanocomposites. Qiu and Netravali [188] showed that the inherent thermal stability of nanofillers could be the main reason for improving thermal stability of nanocomposites. Their TGA results indicated that PVA had two decomposition temperatures detected at 266 and 276 °C. These temperatures were increased by 20 and 29 °C, respectively, for PVA/10 wt% HNT nanocomposites and then a further increase was reported by 29 and 67 °C accordingly for PVA/20 wt% HNT nanocomposites. Moreover, Liu et al. [186] reported that the T_g and T_m of PVA/ NFC nanocomposites were increased dramatically from 77.4 to 83.2 °C and from 288 to 331 °C, respectively, with increasing the NFC content from 0 to 60 wt%. Such a finding arose from strong interfacial bonding between nanofillers and polymer matrices to restrict the mobility of polymeric chains and further enhance their thermal stability. These results seemed to be completely opposite to those of Asad et al. [183] in terms of slight reduction in T_g and T_m of PVA/oxidised nanocellulose composite when compared with those of neat PVA. As such, it was associated with relatively low thermal stability of fillers particularly at high temperature levels.

Most nanocomposites are transparent materials with the incorporation of a small amount of nanofillers to minimise the light scattering in contrast to conventional composites [182]. In particular, when such nanofillers are well dispersed within polymer matrices, the resulting nanocomposites can possess high optical clarity because of no light scattering points [1]. Zhou et al. [189] reported that there was no significant change in the light transparency of PVA/HNT nanocomposites when compared with that of PVA alone since HNTs were uniformly dispersed within polymer matrices according to the AFM observation. Similar results were confirmed for PVA/NFC nanocomposites by Cai et al. [175]. However, Loryuenyong et al. [114] found that the light transparency of PVA/0.3 wt% GO nanocomposites was decreased by 13.18% relative to that of near PVA alone, and then changed to be completely opaque when increasing the GO content up to 2 wt%. This was attributed to GO aggregation resulting in blocking light paths within polymer matrices. On the other hand, Cano et al. [112] reported that the reduction in light transparency in the UV–visible range of PVA/CNC nanocomposites could be used to protect the products from a light oxidative process for the purpose of food packaging.

Barrier properties of nanocomposites are important for many applications such as protective coating and material packaging particularly when in direct contact with foodstuffs [10, 190]. The content, aspect ratio and dispersion of nanofillers within polymer matrices, as well as their orientation relative to the diffusion direction are deemed as major factors to influence barrier properties of nanocomposites [29, 191]. Nanofillers improve barrier properties of nanocomposites by creating "tortuous" paths within polymer matrices so that permeable molecules have to follow a zigzag pathway with the diffusion delay in nanocomposites, and the reduction of free volumes of polymer matrices makes it inaccessible to permeable molecules [29, 192]. Consequently, the presence of nanofillers with good dispersion in polymer matrices improves barrier properties of nanocomposites by means of permeability reduction [180]. Strawhecker and Manias [181] found that well-dispersed MMTs (4–6 wt%) within PVA matrices gave rise to exfoliated nanocomposite structures with the reduction of WVP by 40% when compared with that of neat PVA. In a similar manner, Feiz and Navarchian [185] found the water vapour transmission rate (WVTR) and WVP of PVA/CsMMT composite hydrogels decreased linearly by 37.5 and 45%, respectively, with increasing the CsMMT content from 1 to 5 wt% relative to those of neat PVA. Moreover, Loryuenyong et al. [114] concluded that the good dispersion of 2 wt% GO sheets into PVA matrices built tortuous paths within nanocomposites, which reduced the WVP and oxygen permeability by 21 and 76%, respectively. Similarly, Aloui et al. [193] reported that the WVP was decreased by 42% in PVA/5 wt% CNC/3 wt% HNT nanocomposites as opposed to that of PVA. It was evidenced from SEM observation by illustrating well-dispersed CNCs and HNTs to generate much longer tortuous paths.

1.3.8 Applications

Reinforcing PVA and PVA blends with alternative nanofillers widely expand the range of their applications due to significantly improved mechanical, thermal, barrier, optical, electrical and magnetic properties [31], as illustrated in Fig. 1.22. Nanoclays have a wide variety range of applications due to their availability and relatively low cost [194]. For instance, Zhou et al. [189] found that PVA/HNT bionanocomposites had better surface properties like surface chemistry and nanotopography as compared with those of neat PVA leading to their success in drug delivery and bone tissues due to the compatibility with osteoblast and fibroblast cells. On the other hand, Swapna et al. [148] stated that PVA/HNT nanocomposites could be successfully used for flame retardant applications due to improving their decomposition temperatures resulting from high thermal stability of HNTs, as well as their barrier effect against mass and heat transfer. Koosha et al. [184] found that PVA/chitosan/Na-MMT composite nanofibres with exfoliated structures took place when a homogenous distribution of nanoclay layers was detected within polymer matrices. Additionally, porous surfaces with appropriate pore sizes and non-cytotoxicity properties led to 90% viability of human fibroblast cell line to be successfully used for skin tissue engineering. Furthermore, Feiz and Navarchian [185] found that PVA/CsMMT composite hydrogels with intercalated/partially exfoliated structures could be successfully implemented as a wound dressing due to their high hydrophilicity and acceptable water uptake to benefit the absorption of exudative wounds and accelerate the healing process. Whereas, Noshirvani et al. [195] prepared PVA/ST/MMT nanocomposites with an increase in tensile strength by 17.38% and a decrease in WVP by 16.46%, as opposed to those of PVA/ST blends, which could be efficiently used for food packaging applications. Moreover, Yang et al. [196] manufactured electrolyte membrane based on PVA/MMT nanocomposites for direct methanol fuel cell (DMFC) with lower weight losses by 26.41, 18.73 and 17.67% at elevated temperatures of 400, 500 and 600°C since nanocomposites had higher thermal stability than that of PVA membranes. In addition, PVA/MMT nanocomposite membranes had high ionic conductivity and lower methanol permeability in contrast to commercial membranes like Nafion 117.

Moreover, PVA/functional MWCNT (f-MWCNT) nanocomposite membranes were prepared by Youssef et al. [197] for wastewater treatment. Their results showed that the tensile strength of PVA/f-MWCNT nanocomposites was increased linearly by 165.04% with increasing the f-MWCNT content from 0 to 6 wt% relative to that of neat PVA. Additionally, the elongation at break was decreased by 60.24% at the same nanofiller content due to their inherent toughness. Furthermore, the removal efficiency of heavy metals and pesticides from wastewater was in range

Fig. 1.22 Widespread applications of PVA nanocomposites based on the number of research publications on Scopus Database from 1990 to 2019

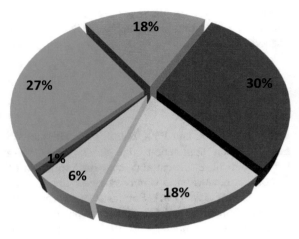

■ **Materials Science**

■ **Chemical applications**

■ **Engineering applications**

   Medical applications

   Agricultural and packaging applications

■ **Other applications**

of 95–100% for PVA/f-MWCNT nanocomposites at the f-MWCNT contents of 6–10 wt%, relative to that for neat PVA. This was because there existed typical antimicrobial activity of nanofillers and occurrence of different reaction rates and reversibility. Moreover, Xue et al. [198] found the agglomeration of MWCNTs was the main reason for limited electrical conductivity of yarns based on PVA/MWCNT nanocomposites. Consequently, PVA/MWCNT nanocomposites were used as a coating material for other yarns like cotton, silk, wool/nylon, polyester and polypropylene so that conductive yarns could be generated with small diameter, low density, high strength and flexibility. Xu et al. [152] revealed that increasing tensile strength and Young's modulus of PVA/3 wt% GO nanocomposites by 69.23 and 128.57%, respectively, could provide competitive applications for structural materials with high strength and flaw tolerance.

1.4   Permeability Modelling of Nanocomposites

The permeation process of gas or liquid molecules through polymer films has four major steps. It starts with the sorption of permeable molecules on the films, and is followed by the dissolution of these molecules inside the films. Then dissolved permeable molecules diffuse in films, and finally are adsorbed on other film surfaces [191]. Consequently, the permeability is completely determined by a diffusion/solubility mechanism that can be explained mathematically using the following equation [180, 191, 199].

$$P_o = D_o \times S_o \qquad (1.1)$$

where P_o, D_o and S_o are the permeability, diffusion and solubility coefficients of polymer matrices in nanocomposites, respectively [191, 199].

This equation can be used to describe the permeability of polymer nanocomposites with the assumption of no voids taking place at polymer/nanofiller interfaces due to strong filler–matrix interfacial bonding [199]. The solubility of permeable molecules in nanocomposites can be given by [180, 191, 199]:

$$S = (1 - \emptyset)S_o \qquad (1.2)$$

where S and S_o are the solubility coefficients in nanocomposites and polymer matrices [191], respectively. ϕ is the volume fraction of nanofillers [199–201] that can be expressed as:

$$\frac{1}{\emptyset} = 1 + \frac{\rho_i(1 - \mu_i)}{\rho_p \mu_i} \qquad (1.3)$$

where ρ_i and ρ_p are the densities of nanofillers and polymer matrices, respectively, and μ_i is the weight fraction of nanofillers [199–201].

Moreover, permeable molecules should follow tortuous paths in nanocomposite films for the diffusion through them. Therefore, the diffusion coefficient (D) of nanocomposites can be written as [180, 191, 199]:

$$D = \frac{D_o}{\tau} \qquad (1.4)$$

where D_o and τ are the diffusion coefficients in polymer matrices and the tortuosity, respectively [191]. The permeability of nanocomposites (P_c) relative to the permeability of polymer matrices can be calculated by combining Eqs. (1.2) and (1.4) below [180, 191, 199]:

$$\frac{P_c}{P_o} = \frac{(1 - \emptyset)}{\tau} \qquad (1.5)$$

Consequently, the relative permeability of nanocomposites depends on volume fraction and tortuosity factor when defined by Nielsen's model as follows [191, 202]:

$$\tau = \frac{l'}{l} \tag{1.6}$$

where l' and l are the distances of zigzag path in nanocomposite films and the straight path in polymer matrices films, respectively [202]. Several models have been developed to calculate τ depending on the content, aspect ratio and geometrical shape of nanofillers [199] according to Table 1.4. In most of these models, it is assumed that nanofillers have regular geometrical shapes in the form of hexagonal flakes, ribbons and disks [199].

Some expressions in relation to τ are used to predict the relative permeability of nanocomposites with the consideration of nanofiller orientation when dispersed within polymer matrices [191], as listed in Table 1.5.

The order parameter (S) provides a good expression for nanofiller orientation particularly with respect to ribbon-like and platelet-like nanofillers as follows [180, 191, 203]:

$$S = \frac{1}{2}\left(3\cos^2\theta - 1\right) \tag{1.7}$$

where θ is the angle between the direction of preferred orientation (n) and normal unit vectors of nanofillers (p) [203]. As such, S has three different values including (i) $S = -1/2$ ($\theta = 90°$) for nanofillers without barrier effect against permeable molecules, (ii) $S = 0$ ($\theta = 54.74°$) for randomly oriented nanofillers and (iii) $S = 1$ ($\theta = 0°$) for nanofillers with the regular arrangement [180, 201, 203], Fig. 1.23.

Tan and Thomas [206] suggested that Nielsen model could be used accurately in terms of both WVP and gas permeability particularly for polymer/clay nanocomposites because of its dependence on the first-order formula when compared with other models. Moreover, Bharadwaj model is used successfully for polymer/clay nanocomposites with incomplete exfoliation, while Cussler models are more applicable for nanocomposites with low volume fraction and aspect ratio of nanofillers [206]. These findings were proven by Saritha et al. [207] when the experimental permeability results of chlorobutyl rubber/modified-MMT (Cloisite-15A) nanocomposites were compared with those obtained from theoretical models. Their results showed that experimentally determined gas permeabilities (O_2, N_2 and CO_2) for chlorobutyl rubber/Cloisite-15A nanocomposites had good agreement with Gusev and Lusti model and Nielsen model, while Cussler model revealed a much closer correlation with experimental results with low aspect ratio and low volume fraction of Cloisite-15A clays. Nielsen model and Cussler model were used by Liu et al. [208] as well for the comparison with the experimental data with respect to gas permeabilities of PVA/GO nanocomposites. Their results revealed that the hydrogen permeabilities of PVA/GO nanocomposites and PVA/modified-GO nanocomposites were decreased in a linear manner by 90 and 94%, respectively, with increasing the nanofiller content from 0 to 3 wt% in contrast to that of neat PVA. Moreover, the experimental data of PVA/GO nanocomposites were placed between Nielsen model and Cussler model while PVA/modified-GO

Table 1.4 Different theoretical models for tortuosity factor

Model	Nanofiller geometry	Tortuosity formula (τ)	Remark	References
Nielsen	Ribbon nanofillers	$1 + \frac{\alpha\phi}{2}$	α: nanofiller aspect ratio	[199, 202]
Cussler	Ribbon nanofillers	$1 + \frac{\alpha^2\phi^2}{4(1-\phi)}$	Perpendicular alignment	[180, 199]
	Ribbon nanofillers	$1 + \frac{\alpha^2\phi^2}{8(1-\phi)}$	Alignment and misalignment	
	Hexagonal nanofillers	$1 + \frac{\alpha^2\phi^2}{54(1-\phi)}$		
Lape and Cussler	Ribbon nanofillers	$\left(1 + \frac{\alpha\phi}{3}\right)^2$		[199]
Maxwell	Spherical nanofillers	$1 + \frac{1+\left(\frac{\phi}{2}\right)}{1-\phi}$		[180]
	Cylindrical nanofillers	$\frac{1+\phi}{1-\phi}$		
Gusev and Lusti	Disk nanofillers	$\exp\left[\left(\frac{\alpha\phi}{3.47}\right)^{0.71}\right]$		[180, 199]
Fredrickson and Bicerano	Disk nanofillers	$4\left[\frac{1+x+0.1245x^2}{2+x}\right]^2$	$x = \frac{\pi\alpha\phi}{2}\ln\frac{\alpha}{2}$	[199]

Table 1.5 Theoretical models of relative permeability

Model	Nanofiller arrangement	Relative permeability formula (P_c/P_o)	References
Nielsen	Regular arrangement	$\frac{1-\phi}{1+\left(\frac{\alpha}{2}\right)\phi}$	[191, 204–206]
	Random arrangement	$\frac{1-\phi}{1+\frac{1}{3}\left(\frac{\alpha}{2}\right)\phi}$	
Cussler	Regular arrangement	$\frac{1-\phi}{1+\left(\frac{\alpha\phi}{2}\right)^2}$	[191, 205]
	Random arrangement	$\frac{1-\phi}{\left(1+\frac{\alpha\phi}{3}\right)^2}$	
Bharadwaj	Any arrangement	$\frac{1-\phi}{1+\frac{2\alpha\phi}{3}\left(S+\frac{1}{2}\right)}$	[191, 206]
Gusev and Lusti	Random arrangement	$\frac{1-\phi}{\exp\left[\left(\frac{\alpha\phi}{3.47}\right)^{0.71}\right]}$	[204, 205]
Fredrickson and Bicerano	Random arrangement	$\frac{1-\phi}{4\left[\frac{1+x+0.1245x^2}{2+x}\right]^2}$	[204–206]

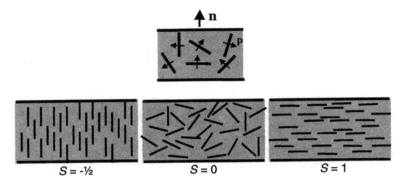

Fig. 1.23 Relationship between nanofiller orientation and order parameters. Image taken from [203] with the copyright permission from American Chemical Society

nanocomposites demonstrated better agreement with Cussler model for well-aligned nanofillers as GO modification improved their overall dispersion within PVA matrices in a nanocomposite system [208].

1.5   Summary

In short, biodegradable polymers have received great attention nowadays to replace petroleum-based polymers due to their availability, renewability and ecofriendly characteristics to tackle critical problems with respect to plastic wastes. PVA is one of these biopolymers in possession of good mechanical properties and acceptable thermal properties with complete transparency, as well as low gas permeability. PVA as a water-soluble polymer has good compatibility with a variety of polymers and nanofillers. PVA can be used as a base polymer or a part of polymer blends in nanocomposite systems. For example, plasticised PVA/ST blends have been studied for decades due to their remarkable blend properties like good biodegradability, cost effectiveness and ductility as opposed to those of neat PVA, as well as high strength and formability relative to those of neat ST. Such blends still have poor barrier properties due to their high water affinity of the components to restrict their applications on a wider scale. The preparation of new nanocomposite systems can overcome such limitations with the inclusion of nanofillers like HNTs with inherent barrier properties and non-toxicity in order to ultimately benefit their food packaging applications.

References

1. Sapalidis AA, Katsaros FK, Kanellopoulos NK (2011) PVA/montmorillonite nanocomposites: development and properties. In: Cuppoletti J (ed) Nanocomposites and polymers with analytical methods. InTechOpen, China, pp 29–50

2. Chiellini E, Corti A, D'Antone S, Solaro R (2003) Biodegradation of poly (vinyl alcohol) based materials. Prog Polym Sci 28:963–1014

3. Baker MI, Walsh SP, Schwartz Z, Boyan BD (2012) A review of polyvinyl alcohol and its uses in cartilage and orthopedic applications. J Biomed Mater Res B Appl Biomater 100 (5):1451–1457

4. Gaaz TS, Sulong AB, Akhtar MN, Kadhum AAH, Mohamad AB, Al-Amiery AA (2015) Properties and applications of polyvinyl alcohol, halloysite nanotubes and their nanocomposites. Molecules 20(12):22833–22847

5. Teodorescu M, Bercea M, Morariu S (2018) Biomaterials of poly(vinyl alcohol) and natural polymers. Polym Rev 58(2):247–287

6. Tânase EE, Popa VI, Popa ME, Râpâ M, Popa O (2016) Biodegradation study of some food packaging biopolymers based on PVA. Bulletin UASVM Animal Sci Biotechnol 73(1):1–5. https://doi.org/10.15835/buasvmcn-asb:11948

7. He Z, Xiong L (2012) Evaluation of physical and biological properties of polyvinyl alcohol/chitosan blend films. J Macromol Sci B 51(9):1705–1714

8. Li HZ, Chen SC, Wang YZ (2014) Thermoplastic PVA/PLA blends with improved processability and hydrophobicity. Ind Eng Chemist Res 53:17355–17361

9. Huang HD, Ren PG, Chen J, Zhang WQ, Ji X, Li ZM (2012) High barrier graphene oxide nanosheet/poly(vinyl alcohol) nanocomposite films. J Mem Sci 409–410:156–163

10. Rhim JW, Ng PKW (2007) Natural biopolymer-based nanocomposite films for packaging applications. Crit Rev Food Sci Nutr 47(4):411–433

11. Mangaraj S, Yadav A, Bal LM, Dash SK, Mahanti NK (2018) Application of biodegradable polymers in food packaging industry: a comprehensive review. J Packag Technol Res 3 (1):77–96

12. Ramos ÓL, Pereira RN, Cerqueira MA, Martins JR, Teixeira JA, Malcata FX, Vicente AA (2018) Bio-based nanocomposites for food packaging and their effect in food quality and safety. In: Grumezescu AM, Holban AM (eds) Food packaging and preservation. Elsevier, London, UK, pp 271–306

13. Rhim JW, Park HM, Ha CS (2013) Bio-nanocomposites for food packaging applications. Prog Polym Sci 38:1629–1652

14. Mensitieri G, Maio ED, Buonocore GG, Nedi I, Oliviero M, Sansone L, Iannace S (2011) Processing and shelf life issues of selected food packaging materials and structures from renewable resources. Trend Food Sc Technol 22:72–80

15. Sorrentino A, Gorrasi G, Vittoria V (2007) Potential perspectives of bio-nanocomposites for food packaging applications. Trend Food Sci Technol 18(2):84–95

16. Arora A, Padua GW (2010) Review: nanocomposites in food packaging. J Food Sci 75 (1):43–49

17. Souza VGL, Fernando AL (2016) Nanoparticles in food packaging: biodegradability and potential migration to food—a review. Food Packag Shelf Life 8:63–70

18. Cerqueira MA, Vicente AA, Pastrana LM (2018) Nanotechnology in food packaging: opportunities and challenges. In: Cerqueira MAPR, Lagaron JM, Castro LMP, Vicente AAMOS (ed) Nanomaterials for food packaging: materials, processing technologies, and safety issues. Elesvier, India, pp 1–11

19. Youssef AM, El-Sayed SM (2018) Bionanocomposites materials for food packaging applications: concepts and future outlook. Carbohydr Polym 193:19–27

20. Duncan TV (2011) Applications of nanotechnology in food packaging and food safety: barrier materials, antimicrobials and sensors. J Colloid Interf Sci 363(1):1–24

21. Shah AA, Hasan F, Hameed A, Ahmed S (2008) Biological degradation of plastics: a comprehensive review. Biotechnol Adv 26(3):246–265
22. Avérous L, Pollet E (2012) Biodegradable polymers. In: Avérous L, Pollet E (eds) Environmental silicate nano-biocomposites. Springer, London, pp 13–39
23. Chandra R, Rustgi R (1998) Biodegradable polymers. Prog Polym Sci 23:1273–1335
24. Lucas N, Bienaime C, Belloy C, Queneudec M, Silvestre F, Nava-Saucedo JE (2008) Polymer biodegradation: mechanisms and estimation techniques. Chemosphere 73(4):429–442
25. Siracusa V, Rocculi P, Romani S, Rosa MD (2008) Biodegradable polymers for food packaging: a review. Trend Food Sci Technol 19:634–643
26. Averous L, Boquillon N (2004) Biocomposites based on plasticized starch: thermal and mechanical behaviours. Carbohydr Polym 56:111–122
27. Hu D, Wang L (2016) Fabrication of antibacterial blend film from poly (vinyl alcohol) and quaternized chitosan for packaging. Mater Res Bulletin 78:46–52
28. Saxena SK (2004) Polyvinyl alcohol (PVA)—chemical and technical assessment (CTA). JECFA 61:1–3
29. Ray SS, Bousmina M (2005) Biodegradable polymers and their layered silicate nanocomposites: in greening the 21st century materials world. Prog Mater Sci 50:962–1079
30. Mohsin M, Hossin A, Haik Y (2011) Thermomechanical properties of poly(vinyl alcohol) plasticized with varying ratios of sorbitol. Mater Sci Eng A 528:925–930
31. Mousa MH, Dong Y, Davies IJ (2016) Recent advances in bionanocomposites: preparation, properties, and applications. Inter J Polymer Mater Polymer Biomater 65(5):225–254
32. Tang X, Alavi S (2011) Recent advances in starch, polyvinyl alcohol based polymer blends, nanocomposites and their biodegradability. Carbohydr Polymer 85:7–16
33. Jang J, Lee DK (2003) Plasticizer effect on the melting and crystallization behavior of polyvinyl alcohol. Polymer 44:8139–8146
34. Sreedhar B, Chattopadhyay DK, Karunakar MSH, Sastry ARK (2006) Thermal and surface characterization of plasticized starch polyvinyl alcohol blends crosslinked with epichlorohydrin. J Appl Polym Sci 101(1):25–34
35. Zhang Y, Han JH (2006) Mechanical and thermal characteristics of pea starch films plasticized with monosaccharides and polyols. J Food Sci 71(2):109–118
36. Roohani M, Habibi Y, Belgacem NM, Ebrahim G, Karimi AN, Dufresne A (2005) Cellulose whiskers reinforced polyvinyl alcohol copolymers nanocomposites. Eur Polym J 44:2489–2498
37. Mishra R, Rao KJ (1999) On the formation of poly(ethylenoxide)-poly(vinylalcohol) blends. Eur Polym J 35:1883–1894
38. Gohil JM, Bhattacharya A, Ray P (2006) Studies on the cross-linking of poly(vinyl alcohol). J Polym Res 13:161–169
39. García NL, Famá L, D'Accorso NB, Goyanes S (2015) Biodegradable starch nanocomposites. In: Thakur VK, Thakur MK (eds) Eco-friendly polymer nanocomposites: processing and properties. Springer India, New Delhi, pp 17–77
40. Corre DL, Bras J, Dufresne A (2010) Starch nanoparticles: a review. Biomacromol 11:1139–1153
41. Whistler RL, Daniel J (2000) Starch. In: Othmer K (ed) Encyclopedia of chemical technology. Wiley, pp 1–18
42. Visakh PM (2015) Starch: state-of-the-art, new challenges and opportunities. In: Visakh PM, Yu L (ed) Starch-based blends, composites and nanocomposites. Royal Society of Chemistry. pp 1–16
43. Shi R, Zhang Z, Liu Q, Han Y, Zhang L, Chen D, Tian W (2007) Characterization of citric acid/glycerol co-plasticized thermoplastic starch prepared by melt blending. Carbohydr Polym 69(4):748–755
44. Avérous L, Halley PJ (2009) Biocomposites based on plasticized starch. Biofuels Bioprod Bioref 3(3):329–343

45. Robyt JF (2008) Starch: structure, properties, chemistry, and enzymology. In: Fraser-Reid B, Tatsuta K, Thiem J (ed) Glycoscience: chemistry and chemical biology. Springer, Berlin, Heidelberg, pp 1437–1472

46. Majdzadeh-Ardakani K, Navarchian AH, Sadeghi F (2010) Optimization of mechanical properties of thermoplastic starch/clay nanocomposites. Carbohydr Polym 79(3):547–554

47. Chiou BS, Yee E, Glenn GM, Orts WJ (2005) Rheology of starch–clay nanocomposites. Carbohydr Polym 59(4):467–475

48. Jose J, Al-Harthi MA, Al-Ma'adeed MA, Dakua JB, De SK (2015) Effect of graphene loading on thermomechanical properties of poly(vinyl alcohol)/starch blend. J App Polym Science 132(16):41827

49. Rahmat AR, Rahman WAWA, Sin LT, Yussuf AA (2009) Approaches to improve compatibility of starch filled polymer system: a review. Mater Sci Eng C 29(8):2370–2377

50. Liu Z, Feng Y, Yi XS (1999) Thermoplastic starch/PVAl compounds: preparation, processing, and properties. J Appl Polym Sci 74:2667–2673

51. Luo X, Li J, Lin X (2012) Effect of gelatinization and additives on morphology and thermal behavior of corn starch/PVA blend films. Carbohydr Polym 90(4):1595–1600

52. Tănase EE, Popa ME, Rapa M, Popa O (2015) Preparation and characterization of biopolymer blends based on polyvinyl alcohol and starch. Roman Biotechnol Lett 20 (20):10306–10315

53. Wang W, Zhang H, Dai Y, Hou H, Dong H (2015) Effects of low poly(vinyl alcohol) content on properties of biodegradable blowing films based on two modified starches. J Thermoplas Compos Mater 30(7):1017–1030

54. Sin LT, Rahman WAWA, Rahmat AR, Khan MI (2010) Detection of synergistic interactions of polyvinyl alcohol–cassava starch blends through DSC. Carbohydr Polym 79(1):224–226

55. Rahman WAWA, Sin LT, Rahmat AR, Samad AA (2010) Thermal behaviour and interactions of cassava starch filled with glycerol plasticized polyvinyl alcohol blends. Carbohydr Polym 81:805–810

56. Zanela J, Olivato JB, Dias AP, Grossmann MVE, Yamashita F (2015) Mixture design applied for the development of films based on starch, polyvinyl alcohol, and glycerol. J App Polym Science 132(43):42697

57. Castro-Aguirre E, Iñiguez-Franco F, Samsudin H, Fang X, Auras R (2016) Poly(lactic acid)-Mass production, processing, industrial applications, and end of life. Adv Drug Deliv Rev 107:333–366

58. Avérous L (2004) Biodegradable multiphase systems based on plasticized starch: a review. J Macromol Sci C 44 (3):231–274

59. Zhang R, Xu W, Jiang F (2012) Fabrication and characterization of dense Chitosan/polyvinyl-alcohol/poly-lactic-acid blend membranes. Fiber Polym 13(5):571–575

60. Liu Y., Wei H, Wang Z, Li Q, Tian N (2018) Simultaneous enhancement of strength and toughness of PLA induced by miscibility variation with PVA. Polym 10(10):1178

61. Shuai X, He Y, Askawa N, Inoue Y (2000) Miscibility and phase structure of binary blends of poly(L-lactide) and poly(vinyl alcohol). J Appl Polym Sci 81:762–772

62. Restrepo I, Medina C, Meruane V, Akbari-Fakhrabadi A, Flores P, Rodríguez-Llamazares S (2018) The effect of molecular weight and hydrolysis degree of poly(vinyl alcohol) (PVA) on the thermal and mechanical properties of poly(lactic acid)/PVA blends. Polímeros 28(2):169–177

63. Yeh JT, Yang MC, Wu CJ, Wu X, Wu CS (2008) Study on the crystallization kinetic and characterization of poly(lactic acid) and poly(vinyl alcohol) blends. Polym Plas Technol Eng 47(12):1289–1296

64. Hu Y, Wang Q, Tang M (2013) Preparation and properties of starch-g-PLA/poly(vinyl alcohol) composite film. Carbohydr Polym 96(2):384–388

65. Van den Broek LAM, Knoop RJI, Kappen FHJ, Boeriu CG (2015) Chitosan films and blends for packaging material. Carbohydr Polym 116:237–242

66. Aoi K, Takasu A, Okada M (1995) New chitin-based polymer hybrids, 1. Miscibility of poly (vinyl1 alcohol) with chitin derivatives. Macromol Rapid Commun 16:757–761

67. Takasu A, Aoi K, Tsuchiya M, Okada M (1998) New chitin-based polymer hybrids, 4. Soil burial degradation behavior of poly(vinyl alcohol)/chitin derivative miscible blends. J App Polym Sci 73:1171–1179

68. Mujtaba M, Morsi RE, Kerch G, Elsabee MZ, Kaya M, Labidi J, Khawar KM (2019) Current advancements in chitosan-based film production for food technology; a review. Int J Biol Macromol 121:889–904

69. Kasai D, Chougale R, Masti S, Chalannavar R, Malabadi RB, Gani R (2018) Influence of syzygium cumini leaves extract on morphological, thermal, mechanical, and antimicrobial properties of PVA and PVA/chitosan blend films. J App Polym Sci 135(17):46188

70. Giannakas A, Vlacha M, Salmas C, Leontiou A, Katapodis P, Stamatis H, Barkoula NM, Ladavos A (2016) Preparation, characterization, mechanical, barrier and antimicrobial properties of chitosan/PVOH/clay nanocomposites. Carbohydr Polym 140:408–415

71. Tripathi S, Mehrotra GK, Dutta PK (2009) Physicochemical and bioactivity of cross-linked chitosan-PVA film for food packaging applications. Int J Biol Macromol 45(4):372–376

72. Liu Y, Wang S, Lan W (2018) Fabrication of antibacterial chitosan-PVA blended film using electrospray technique for food packaging applications. Int J Biol Macromol 107:848–854

73. Djagny KB, Wang Z, Xu S (2001) Gelatin: a valuable protein for food and pharmaceutical industries: review. Crit Rev Food Sci Nutr 41(6):481–492

74. Devi N, Sarmah M, Khatun B, Maji TK (2017) Encapsulation of active ingredients in polysaccharide-protein complex coacervates. Adv Colloi Interf Sci 239:136–145

75. Pawde SM, Deshmukh K (2008) Characterization of polyvinyl alcohol/gelatin blend hydrogel films for biomedical applications. J Appl Polym Sci 109(5):3431–3437

76. Pal K, Banthia AK, Majumdar DK (2007) Preparation and characterization of polyvinyl alcohol-gelatin hydrogel membranes for biomedical applications. AAPS Pharm Sci Tech 8 (1):E1–E5

77. Liu Y, Geever LM, Kennedy JE, Higginbotham CL, Cahill PA, McGuinness GB (2010) Thermal behavior and mechanical properties of physically crosslinked PVA/gelatin hydrogels. J Mech Beh Biomed Mater 3(2):203–209

78. Ino JM, Sju E, Ollivier V, Yim EKF, Letourneur D, Visage CL (2013) Evaluation of hemocompatibility and endothelialization of hybrid poly(vinyl alcohol) (PVA)/gelatin polymer films. J Biomed Mater Res B Appl Biomater 101(8):1549–1559

79. Pawde SM, Deshmukh K, Parab S (2008) Preparation and characterization of poly(vinyl alcohol) and gelatin blend films. J Appl Polym Sci 109(2):1328–1337

80. Gao X, Tang K, Liu J, Zheng X, Zhang Y (2014) Compatibility and properties of biodegradable blend films with gelatin and poly(vinyl alcohol). J Wuhan Univ Technol Mater Sci Ed 29(2):351–356

81. Jayasekara R, Harding I, Bowater I, Christie GBY, Lonergan GT (2004) Preparation, surface modification and characterisation of solution cast starch PVA blended films. Polym Test 23 (1):17–27

82. Zhou J, Ma Y, Ren L, Tong J, Liu Z, Xie L (2009) Preparation and characterization of surface crosslinked TPS/PVA blend films. Carbohydr Polym 76(4):632–638

83. Mohsin M, Hossin A, Haik Y (2011) Thermal and mechanical properties of poly(vinyl alcohol) plasticized with glycerol. J Appl Polym Sci 122(5):3102–3109

84. Zou GX, Ping-Qu J, Liang-Zou X (2008) Extruded starch/PVA composites: water resistance, thermal properties, and morphology. J Elastom Plast 40(4):303–316

85. Zou GX, Qu JP, Zou XL (2007) Optimization of water absorption of starch/PVA composites. Polym Compos 28(5):674–679

86. Lim LY, Wan LSC (2008) The effect of plasticizers on the properties of polyvinyl alcohol films. Drug Devel Ind Pharm 20(6):1007–1020

87. Kopčilová M, Hubáčková J, Růžička J, Dvořáčková M, Julinová M, Koutný M, Tomalová M, Alexy P, Bugaj P, Filip J (2012) Biodegradability and mechanical properties of poly (vinyl alcohol)-based blend plastics prepared through extrusion method. J Polym Environ 21 (1):88–94

88. Lim M, Kwon H, Kim D, Seo J, Han H, Khan SB (2015) Highly-enhanced water resistant and oxygen barrier properties of cross-linked poly(vinyl alcohol) hybrid films for packaging applications. Prog Organic Coating 85:68–75

89. Ismail H, Zaaba NF (2011) Effect of additives on properties of polyvinyl alcohol (PVA)/tapioca starch biodegradable films. Polym Plast Technol Eng 50(12):1214–1219

90. Aydın AA, Ilberg V (2016) Effect of different polyol-based plasticizers on thermal properties of polyvinyl alcohol: starch blends. Carbohydr Polym 136:441–448

91. Shi R, Bi J, Zhang Z, Zhu A, Chen D, Zhou X, Zhang L, Tian W (2008) The effect of citric acid on the structural properties and cytotoxicity of the polyvinyl alcohol/starch films when molding at high temperature. Carbohydr Polym 74(4):763–770

92. Tudorachi N, Cascaval CN, Rusu M, Pruteanu M (2000) Testing of polyvinyl alcohol and starch mixtures as biodegradable polymeric materials. Polym Test 19:785–799

93. Das K, Ray D, Bandyopadhyay NR, Gupta A, Sengupta S, Sahoo S, Mohanty A, Misra M (2010) Preparation and characterization of cross-linked starch/poly(vinyl alcohol) green films with low moisture absorption. Ind Eng Chemist Res 49:2176–2185

94. Ramaraj B (2007) Crosslinked poly(vinyl alcohol) and starch composite films: study of their physicomechanical, thermal, and swelling properties. J App PolyM Sci 103(2):1127–1132

95. Ramaraj B (2007) Crosslinked poly(vinyl alcohol) and starch composite films. II. Physicomechanical, thermal properties and swelling studies. J Appl Polym Sci 103(2):909–916

96. Yoon SD, Chough SH, Park HR (2006) Properties of starch-based blend films using citric acid as additive. II. J App Polym Sci 100(3):2554–2560

97. Yoon SD, Chough SH, Park HR (2007) Preparation of resistant starch/poly(vinyl alcohol) blend films with added plasticizer and crosslinking agents. J App Polym Sci 106(4):2485–2493

98. Yin Y, Li J, Liu Y, Li Z (2005) Starch crosslinked with poly(vinyl alcohol) by boric acid. J App Polym Sci 96(4):1394–1397

99. Mao L, Imam S, Gordon S, Cinelli P, Chiellini E (2002) Extruded cornstarch–glycerol–polyvinyl alcohol blends: mechanical properties, morphology, and biodegradability. J Polym Environ 8(4):205–211

100. Sin LT, Rahmat AR, Rahman WAWA, Sun ZY, Samad AA (2010) Rheology and thermal transition state of polyvinyl alcohol–cassava starch blends. Carbohydr Polym 81(3):737–739

101. Zanela J, Bilck AP, Casagrande M, Grossmann MVE, Yamashita F (2018) Polyvinyl alcohol (PVA) molecular weight and extrusion temperature in starch/PVA biodegradable sheets. Polímeros 28(3):256–265

102. Chai WL, Chow JD, Chen CC, Chuang FS, Lu WC (2009) Evaluation of the biodegradability of polyvinyl alcohol/starch blends: a methodological comparison of environmentally friendly materials. J Polym Environ 17(2):71–82

103. Tian H, Yan J, Rajulu AV, Xiang A, Luo X (2017) Fabrication and properties of polyvinyl alcohol/starch blend films: effect of composition and humidity. Int J Biol Macromol 96:518–523

104. Gajria AM, Dave V, Gross RA, McCarthy SP (1996) Miscibility and biodegradability of blends of poly(lactic acid) and poly(vinyl acetate). Polym 37(3):437–444

105. Grande R, Carvalho AJF (2011) Compatible ternary blends of chitosan/poly(vinyl alcohol)/poly(lactic acid) produced by oil-in-water emulsion processing. Biomacromol 12(4):907–914

106. Aoi K, Takasu A, Okada A (1997) New chitin-based polymer hybrids. 2. Improved miscibility of chitin derivatives having monodisperse poly(2-methyl-2-oxazoline) side chains with poly(vinyl chloride) and poly(vinyl alcohol). Macromol Rapid Commun 30:6134–6138

107. Maria TMC, Carvalho RA, Sobral PJA, Habitante AMBQ, Solorza-Feria J (2008) The effect of the degree of hydrolysis of the PVA and the plasticizer concentration on the color, opacity, and thermal and mechanical properties of films based on PVA and gelatin blends. J Food Eng 87(2):191–199

108. Mendieta-Taboada O, Sobral PJA, Carvalho RA, Habitante AMBQ (2008) Thermomechanical properties of biodegradable films based on blends of gelatin and poly (vinyl alcohol). Food Hydrocolloid 22(8):1485–1492
109. Gupta B, Agarwal R, Alam MS (2013) Preparation and characterization of polyvinyl alcohol-polyethylene oxide-carboxymethyl cellulose blend membranes. J App Polym Sci 127(2):1301–1308
110. Cano AI, Cháfer M, Chiralt A, González-Martínez C (2015) Physical and microstructural properties of biodegradable films based on pea starch and PVA. J Food Eng 167:59–64
111. Cano AI, Cháfer M, Chiralt A, González-Martínez C (2016) Biodegradation behavior of starch-PVA films as affected by the incorporation of different antimicrobials. Polym Deg Stab 132:11–20
112. Cano A, Fortunati E, Chafer M, Gonzalez-Martınez C, Chiralt A, Kenny JM (2015) Effect of cellulose nanocrystals on the properties of pea starch–poly(vinyl alcohol) blend films. J Mater Sci 50(21):6979–6992
113. Grande R, Pessan LA, Carvalho AJF (2015) Ternary melt blends of poly(lactic acid)/poly (vinyl alcohol)-chitosan. Ind Crop Prod 72:159–165
114. Loryuenyong V, Saewong C, Aranchaiya C, Buasri A (2015) The improvement in mechanical and barrier properties of poly(vinyl alcohol)/graphene oxide packaging films. Packag Technol Sci 28(11):939–947
115. Chen Y, Cao X, Chang PR, Huneault MA (2008) Comparative study on the films of poly (vinyl alcohol)/pea starch nanocrystals and poly(vinyl alcohol)/native pea starch. Carbohydr Polymer 73(1):8–17
116. Azahari NA, Othman N, Ismail H (2011) Biodegradation studies of polyvinyl alcohol/corn starch blend films in solid and solution media. J Phys Sci 22(2):15–31
117. Holland BJ, Hay JN (2001) The thermal degradation of poly(vinyl alcohol). Polym 42:6775–6783
118. Bin-Dahman OA, Jose J, Al-Harthi MA (2016) Effect of natural weather aging on the properties of poly(vinyl alcohol)/starch/graphene nanocomposite. Starch 69(1600005):1–8
119. Spiridon I, Popescu MC, Bodarlau R, Vasile C (2008) Enzymatic degradation of some nanocomposites of poly(vinyl alcohol) with starch. Polym Degrad Stab 93(10):1884–1890
120. Bertuzzi MA, Vidaurre EFC, Armada M, Gottifredi JC (2007) Water vapor permeability of edible starch based films. J Food Eng 80(3):972–978
121. Van de Velde K, Kiekens P (2002) Biopolymers: overview of several properties and consequences on their applications. Polym Test 21:433–442
122. Kamoun EA, Chen X, Eldin MSM, Kenawy ES (2015) Crosslinked poly(vinyl alcohol) hydrogels for wound dressing applications: a review of remarkably blended polymers. Arabian J Chem 8(1):1–14
123. Rafique A, Zia MK, Zuber M, Tabasum S, Rehman S (2016) Chitosan functionalized poly (vinyl alcohol) for prospects biomedical and industrial applications: a review. Int J Biol Macromol 87:141–154
124. Liu R, Xu X, Zhuang X, Cheng B (2014) Solution blowing of chitosan/PVA hydrogel nanofiber mats. Carbohydr Polym 101:1116–1121
125. Zhao L, Mitomo H, Zhai M, Yoshii F, Nagasawa N, Kume T (2003) Synthesis of antibacterial PVA/CM-chitosan blend hydrogels with electron beam irradiation. Carbohydr Polym 53(4):439–446
126. Costa-Júnior ES, Barbosa-Stancioli EF, Mansur AAP, Vasconcelos WL, Mansur HS (2009) Preparation and characterization of chitosan/poly(vinyl alcohol) chemically crosslinked blends for biomedical applications. Carbohydr Polym 76(3):472–481
127. Zhuang PY, Li YL, Fan L, Lin J, Hu QL (2012) Modification of chitosan membrane with poly(vinyl alcohol) and biocompatibility evaluation. Int J Biol Macromol 50(3):658–663
128. Bonilla J, Fortunati E, Atarés L, Chiralt A, Kenny JM (2014) Physical, structural and antimicrobial properties of poly vinyl alcohol–chitosan biodegradable films. Food Hydrocoll 35:463–470

129. Gao X, Tang K, Liu J, Zhang Y (2012) Compatibility of biodegradable composites with gelatine and poly(vinyl alcohol). Paper presented at the 4th international conference on advanced materials and systems. China

130. Hago EE, Li X (2013) Interpenetrating polymer network hydrogels based on gelatin and PVA by biocompatible approaches: synthesis and characterization. Adv Mater Sci Eng 2013:328763

131. Wang M, Li Y, Wu J, Xu F, Zuo Y, Jansen JA (2008) In vitro and in vivo study to the biocompatibility and biodegradation of hydroxyapatite/poly(vinyl alcohol)/gelatin composite. J Biomed Mater Res Part A 85(2):418–426

132. Fan L, Yang H, Yang J, Peng M, Hu J (2016) Preparation and characterization of chitosan/gelatin/PVA hydrogel for wound dressings. Carbohydr Polym 146:427–434

133. Youssef AM (2013) Polymer nanocomposites as a new trend for packaging applications. Polym Plast Technol Eng 52(7):635–660

134. Julkapli NM, Bagheri S, Sapuan SM (2015) Multifunctionalized carbon nanotubes polymer composites: properties and applications. In: Thakur VK, Thakur MK (eds) Eco-friendly polymer nanocomposites-processing and properties, vol 75. Springer, India, New Delhi, pp 155–214

135. Dong P, Prasanth R, Xu F, Wang X, Li B, Shankar R (2015) Eco-friendly polymer nanocomposite-properties and processing. In: Thakur VK, Thakur MK (eds) Eco-friendly polymer nanocomposites-processing and properties, vol 75. Springer, India, New Delhi, pp 1–15

136. Ray S, Quek SY, Easteal A, Chen XD (2006) The potential use of polymer-clay nanocomposites in food packaging. Int J Food Eng 2(4):5

137. Li Y, Tian H, Jia Q, Niu P, Xiang A, Liu D, Qin Y (2015) Development of polyvinyl alcohol/intercalated MMT composite foams fabricated by melt extrusion. J Appl Polym Sci 132(43):42706

138. Majdzadeh-Ardakani K, Nazari B (2010) Improving the mechanical properties of thermoplastic starch/poly(vinyl alcohol)/clay nanocomposites. Compos Sci Technol 70 (10):1557–1563

139. Joussein E, Petit S, Churchman J, Theng B, Richi D, Delvaux B (2005) Halloysite clay minerals—a review. Clay Miner 40:383–426

140. Khoo WS, Ismail H, Ariffin A (2011) Tensile and swelling properties of polyvinyl alcohol/chitosan/halloysite nanotubes nanocomposite. Paper presented at the national postgraduate conference, IEEE, Kuala Lumpur, Malaysia, 19–20 Sept 2011

141. Rawtani D, Agrawal YK (2012) Multifarious applications of halloysite nanotubes: a review. Rev Adv Mater Sci 30:282–295

142. Tully J, Fakhrullin R, Lvov Y (2015) Halloysite clay nanotube composites with sustained release of chemicals. In: Bardosova M, Wagner T (eds) Nanomaterials and nanoarchitectures. Springer, Netherlands, pp 87–118

143. Zhang Y, Tang A, Yang H, Ouyang J (2016) Applications and interfaces of halloysite nanocomposites. App Clay Sci 119:8–17

144. Goda ES, Yoon KR, El-sayed SH, Hong SE (2018) Halloysite nanotubes as smart flame retardant and economic reinforcing materials: a review. Thermochim Acta 669:173–184

145. Darie-Niţă RN, Vasile C (2018) Halloysite containing composites for food packaging applications. In: Kozlowski MA, Spizzirri UG, Cirillo G (eds) Composites materials for food packaging. John Wiley & Sons, USA, pp 73–122

146. Satish S, Tharmavaram M, Rawtani D (2019) Halloysite nanotubes as a nature's boon for biomedical applications. Nanobiomedicine 6:1–16

147. Liu M, Fakhrullin R, Novikov A, Panchal A, Lvov Y (2019) Tubule nanoclay-organic heterostructures for biomedical applications. Macromol Biosci 19(4):1800419

148. Swapna VP, Selvin TP, Suresh KI, Saranya V, Rahana MP, Stephen R (2015) Thermal properties of poly (vinyl alcohol)(PVA)/halloysite nanotubes reinforced nanocomposites. Int J Plast Technol 19(1):124–136

149. Gao W (2015) Synthesis, structure, and characterizations. In: Gao W (ed) Graphene oxide-reduction recipes, spectroscopy, and applications. Springer, Switzerland, pp 1–28
150. McDonald MP, Morozov Y, Hodak JH, Kuno M (2015) Spectroscopy and microscopy of graphene oxide and reduced graphene oxide. In: Gao W (ed) Graphene oxide- reduction recipes, spectroscopy, and applications. Springer, Switzerland, pp 29–60
151. Kim HM, Lee JK, Lee HS (2011) Transparent and high gas barrier films based on poly(vinyl alcohol)/graphene oxide composites. Thin Solid Films 519(22):7766–7771
152. Xu Y, Hong W, Bai H, Li C, Shi G (2009) Strong and ductile poly(vinyl alcohol)/graphene oxide composite films with a layered structure. Carbon 47(15):3538–3543
153. Zhu Y, Murali S, Cai W, Li X, Suk JW, Potts JR, Ruoff RS (2010) Graphene and graphene oxide: synthesis, properties, and applications. Adv Mater 22(35):3906–3924
154. Moon IK, Lee J, Ruoff RS, Lee H (2010) Reduced graphene oxide by chemical graphitization. Nat Commun 1:73
155. Pei S, Cheng HM (2012) The reduction of graphene oxide. Carbon 50:3210–3228
156. Aslam M, Kalyar MA, Raza ZA (2018) Investigation of structural and thermal properties of distinct nanofillers-doped PVA composite films. Polym Bulletin 76(1):73–86
157. Sellam C, Zhai Z, Zahabi H, Picot OT, Deng H, Fu Q, Bilotti E, Peijs T (2015) High mechanical reinforcing efficiency of layered poly(vinyl alcohol)-graphene oxide nanocomposites. Nanocompos 1(2):89–95
158. Liu D, Bian Q, Li Y, Wang Y, Xiang A, Tian H (2016) Effect of oxidation degrees of graphene oxide on the structure and properties of poly (vinyl alcohol) composite films. Compos Sci Technol 129:146–152
159. Lemes AP, Montanheiro TLA, Passador FR, Durán N (2015) Nanocomposites of polyhydroxyalkanoates reinforced with carbon nanotubes: chemical and biological properties. In: Thakur VK, Thakur MK (eds) Eco-friendly polymer nanocomposites: processing and properties. New Delhi, Springer, India, pp 79–108
160. Rafique I, Kausar A, Anwar Z, Muhammad B (2015) Exploration of epoxy resins, hardening systems, and epoxy/carbon nanotube composite designed for high performance materials: a review. Polym Plast Technol Eng 55(3):312–333
161. Myhra S, Riviere JC (2013) Characterisations of nanostructures. Taylor and Francis group, London
162. Karthik PS, Himaja AL, Singh SP (2014) Carbon-allotropes: synthesis methods, applications and future perspectives. Carbon letter 15(4):219–237
163. Kausar A (2018) Eco-polymer and carbon nanotube composite: safe technology. In: Martinez LMT (ed) Handbook of ecomaterials. Springer International Publishing, pp 1–16
164. Paiva MC, Zhou B, Fernando KAS, Lin Y, Kennedy JM, Sun YP (2004) Mechanical and morphological characterization of polymer–carbon nanocomposites from functionalized carbon nanotubes. Carbon 42(14):2849–2854
165. Lin Y, Zhou B, Fernando KAS, Liu P, Allard LF, Sun YP (2003) Polymeric carbon nanocomposites from carbon nanotubes functionalized with matrix polymer. Macromol 36:7199–7204
166. Liu L, Barber AH, Nuriel S, Wargner HD (2005) Mechanical properties of functionlised single-walled carbon nantubes/poly(vinyl alcohol) nanocomposites. Adv Funct Mater 15:975–980
167. Basiuk EV, Anis A, Bandyopadhyay S, Alvarez-Zauco E, Chan SLI, Basiuk VA (2009) Poly (vinyl alcohol)/CNT composites: an effect of cross-linking with glutaraldehyde. Superlattic Microstruct 46(1–2):379–383
168. Wongon J, Thumsorn S, Srisawat N (2016) Poly(vinyl alcohol)/multiwalled carbon nanotubes composite nanofiber. Energy Procedia 89:313–317
169. Ryan KP, Cadek M, Nicolosi V, Blond D, Ruether M, Armstrong G, Swan H, Fonseca A, Nagy JB, Maser WK, Blau WJ, Coleman JN (2007) Carbon nanotubes for reinforcement of plastics? A case study with poly(vinyl alcohol). Compos Sci Technol 67(7–8):1640–1649
170. Moon RJ, Martini A, Nairn J, Simonsen J, Youngblood J (2011) Cellulose nanomaterials review: structure, properties and nanocomposites. Chem Soci Rev 40(7):3941–3994

171. George J, Sabapathi SN, Siddaramaiah (2015) Water soluble polymer-based nanocomposites containing cellulose nanocrystals. In: Thakur VK, Thakur MK (eds) Eco-friendly polymer nanocomposites: processing and properties. Springer India, New Delhi, pp 259–293

172. Muhamad II, Salehudin MH, Salleh E (2015) Cellulose nanofiber for eco-friendly polymer nanocomposites. In: Thakur VK, Thakur MK (eds) Eco-friendly polymer nanocomposites: processing and properties. Springer India, New Delhi, pp 323–365

173. Douglass EF, Avci H, Boy R, Rojas OJ, Kotek R (2017) A review of cellulose and cellulose blends for preparation of bio-derived and conventional membranes, nanostructured thin films, and composites. Polym Rev 58(1):102–163

174. Ibrahim MM, El-Zawawy WK, Nassar MA (2010) Synthesis and characterization of polyvinyl alcohol/nanospherical cellulose particle films. Carbohydr Polym 79(3):694–699

175. Cai J, Chen J, Zhang Q, Lei M, He J, Xiao A, Ma C, Li S, Xiong H (2016) Well-aligned cellulose nanofiber-reinforced polyvinyl alcohol composite film: mechanical and optical properties. Carbohydr Polym 140:238–245

176. Spagnol C, Fragal EH, Witt MA, Follmann HDM, Silva R, Rubira AF (2018) Mechanically improved polyvinyl alcohol-composite films using modified cellulose nanowhiskers as nano-reinforcement. Carbohydr Polym 191:25–34

177. Popescu MC (2017) Structure and sorption properties of CNC reinforced PVA films. Int J Biol Macromol 101:783–790

178. Cheikh SB, Cheikh RB, Cunha E, Lopes PE, Paiva MC (2018) Production of cellulose nanofibers from Alfa grass and application as reinforcement for polyvinyl alcohol. Plast Rubb Compos 47(7):297–305

179. Voronova MI, Surov OV, Guseinov SS, Barannikov VP, Zakharov AG (2015) Thermal stability of polyvinyl alcohol/nanocrystalline cellulose composites. Carbohydr Polym 130:440–447

180. Cui Y, Kumar S, Konac BR, Houcke D (2015) Gas barrier properties of polymer/clay nanocomposites. RSC Adv 5(78):63669–63690

181. Strawhecker KE, Manias E (2000) Structure and properties of poly(vinyl alcohol)/Na⁺ montmorillonite nanocomposites. Chemist Mater 12:2943–2949

182. Zagho MM, Khader MM (2016) The impact of clay loading on the relative intercalation of poly(vinyl alcohol)-clay composites. J Mater Sci Chem Eng 4:20–31

183. Asad M, Saba N, Asiri AM, Jawaid M, Indarti E, Wanrosli WD (2018) Preparation and characterization of nanocomposite films from oil palm pulp nanocellulose/poly (vinyl alcohol) by casting method. Carbohydra Polym 191:103–111

184. Koosha M, Mirzadeh H, Shokrgozarb MA, Farokhib M (2015) Nanoclay-reinforced electrospun chitosan/PVA nanocomposite nanofibers for biomedical applications. RSC Adv 5(14):10479–10487

185. Feiz S, Navarchian AH (2019) Poly(vinyl alcohol) hydrogel/chitosan-modified clay nanocomposites for wound dressing application and controlled drug release. Macromol Res 27(3):290–300

186. Liu D, Sun X, Maiti S, Tian H, Ma Z (2013) Effects of cellulose nanofibrils on the structure and properties on PVA nanocomposites. Cellulose 20(6):2981–2989

187. Nistor MT, Vasile C (2012) Influence of the nanoparticle type on the thermal decomposition of the green starch/poly(vinyl alcohol)/ montmorillonite nanocomposites. J Thermal Analy Calorim 111(3):1903–1919

188. Qiu K, Netravali AN (2013) Halloysite nanotube reinforced biodegradable nanocomposites using noncrosslinked and malonic acid crosslinked polyvinyl alcohol. Polym Compos 34 (5):799–809

189. Zhou WY, Guo B, Liu M, Liao R, Rabie ABM, Jia D (2010) Poly(vinyl alcohol)/halloysite nanotubes bionanocomposite films: properties and in vitro osteoblasts and fibroblasts response. J Biomed Mater Rese A 93(4):1574–1587

190. Cano A, Fortunati E, Chafer M, Kenny JM, Chiralt A, Gonzalez-Martínez C (2015) Properties and ageing behaviour of pea starch films as affected by blend with poly(vinyl alcohol). Food Hydrocolloids 48:84–93

191. Choudalakis G, Gotsis AD (2009) Permeability of polymer/clay nanocomposites: a review. Eur Polym J 45(4):967–984
192. Bhattacharya M, Biswas S, Bhowmick AK (2011) Permeation characteristics and modeling of barrier properties of multifunctional rubber nanocomposites. Polym 52(7):1562–1576
193. Aloui H, Khwaldia K, Hamdi M, Fortunati E, Kenny JM, Buonocore GG, Lavorgna M (2016) Synergistic effect of halloysite and cellulose nanocrystals on the functional properties of PVA based nanocomposites. ACS Sust Chemist Eng 4(3):794–800
194. Idumah CI, Hassan A, Ogbu J, Ndem JU, Nwuzor IC (2018) Recently emerging advancements in halloysite nanotubes polymer nanocomposites. Compos Interf 26(9):751–824
195. Noshirvani N, Ghanbarzadeh B, Fasihi H, Almasi H (2016) Starch–PVA nanocomposite film incorporated with cellulose nanocrystals and MMT: a comparative study. Int J Food Eng 12(1):37–48
196. Yang CC, Lee YJ, Yang JM (2009) Direct methanol fuel cell (DMFC) based on PVA/MMT composite polymer membranes. J Power Sour 188(1):30–37
197. Youssef AM, El-Naggar ME, Malhat FM, El-Sharkawi HM (2019) Efficient removal of pesticides and heavy metals from wastewater and the antimicrobial activity of f-MWCNTs/PVA nanocomposite film. J Cleaner Prod 206:315–325
198. Xue P, Park KH, Tao XM, Chen W, Cheng XY (2007) Electrically conductive yarns based on PVA/carbon nanotubes. Compos Struct 78(2):271–277
199. Picard E, Vermogen A, Gérard JF, Espuche E (2007) Barrier properties of nylon 6-montmorillonite nanocomposite membranes prepared by melt blending: influence of the clay content and dispersion state consequences on modelling. J Membrane Sci 292(1–2):133–144
200. Chen B, Evans JRG (2006) Nominal and effective volume fractions in polymer-clay nanocomposites. Macromol 39:1790–1796
201. Alexandre B, Langevina D, Médéric P, Aubry T, Couderc H, Nguyen QT, Saiter A, Marais S (2009) Water barrier properties of polyamide 12/montmorillonite nanocomposite membranes: structure and volume fraction effects. J Memb Sci 328(1–2):186–204
202. Nielsen LE (1967) Models for the permeability of filled polymer systems. J Macromol Sci A Chemist 1(5):929–942
203. Bharadwaj RK (2001) Notes. Macromol 34:9189–9192
204. Gusev AA, Lusti HR (2001) Rational design of nanocomposites for barrier applications. Adv Mater 13(21):1641–1643
205. Takahashi S, Goldberg HA, Feeney CA, Karim DP, Farrell M, O'Leary K, Paul DR (2006) Gas barrier properties of butyl rubber/vermiculite nanocomposite coatings. Polymer 47(9):3083–3093
206. Tan B, Thomas NL (2016) A review of the water barrier properties of polymer/clay and polymer/graphene nanocomposites. J Memb Sci 514:595–612
207. Saritha A, Joseph K, Thomas S, Muraleekrishnan R (2012) Chlorobutyl rubber nanocomposites as effective gas and VOC barrier materials. Compos A App Sci Manuf 43(6):864–870
208. Liu H, Bandyopadhyay P, Kim NH, Moon B, Lee JH (2016) Surface modified graphene oxide/poly(vinyl alcohol) composite for enhanced hydrogen gas barrier film. Polym Test 50:49–56

Chapter 2
Materials, Manufacturing Process and Characterisation Methods

Abstract In this chapter, important chemical and physical properties of neat polyvinyl alcohol (PVA), starch (ST) from potatoes, glycerol (GL) and halloysite nanotubes (HNTs) were summarised according to the data sheet given by the material suppliers. Moreover, manufacturing procedures and conditions of neat PVA, PVA/ST blends, PVA/GL blends, PVA/ST/GL blends and PVA/ST/GL/HNT bionanocomposite films at different HNT contents were explained in detail. Furthermore, sophisticated material characterisation techniques in term of morphological structures, mechanical, thermal and barrier properties, as well as water resistance were elaborated with their potential applications as food packaging materials.

Keywords Biopolymer blends · Bionanocomposite films · Material properties · Manufacturing process · Characterisation methods

According to the definition of sustainable packaging [1, 2], sustainable packaging materials should be based on renewable or recycled resources of materials, clean production technologies, safe and healthy features throughout their life cycle, as well as by meeting the market requirements for cost and performance. Moreover, nanotechnology has been attractive for packaging applications due to the creation of new material systems with unique properties, resulting from using nanoscale materials with large surface-area-to-volume ratio, which leads to the improvement of chemical and thermal stabilities, better mechanical and barrier properties, as well as lower density compared with microscale and macroscale materials [3]. Consequently, ecofriendly materials such as PVA, ST, GL and HNTs were used to prepare bionanocomposite films using solution casting method in order to design a sustainable material for food packaging applications with cost effectiveness.

© Springer Nature Singapore Pte Ltd. 2020
Z. W. Abdullah and Y. Dong, *Polyvinyl Alcohol/Halloysite Nanotube Bionanocomposites as Biodegradable Packaging Materials*,
https://doi.org/10.1007/978-981-15-7356-9_2

2.1   Materials

2.1.1   Polyvinyl Alcohol (PVA)

PVA is a popular synthetic water-soluble polymer. PVA has a wide range of applications in biochemical, biomedical, pharmaceutical and packaging sectors because of its good compatibility, water solubility and relatively high biodegradability in some environments like active sludge in spite of its poor thermal stability and barrier properties [4, 5]. Full-hydrolysis PVA with a hydrolysis degree of approximately 99% was used, which was purchased from Sigma-Aldrich Pty. Ltd, Australia with specific material properties listed in Table 2.1.

2.1.2   Potato Starch (ST)

ST is a completely biodegradable polymer being extracted from numerous resources like corn, rice, potato, wheat, barley, pea and so on [6, 7]. It is rarely used as a neat polymer because of high brittleness and closeness between melting and degradation temperatures [8]. ST from potatoes with 100% concentration was also supplied by Sigma-Aldrich Pty. Ltd, Australia with detailed properties given in Table 2.1.

2.1.3   Glycerol (GL)

GL is the best plasticiser and compatible agent that have been used with PVA/ST blends for many decades [9] due to their close solubility parameters, as evidenced by 21.10, 22.50 and 23.40 MPa$^{1/2}$ for PVA, ST and GL, respectively [9]. The presence of GL increases the ductility and compatibility of PVA/ST blends by improving polymeric chain mobility and creating hydrogen bonds with the consumption of hydroxyl groups, respectively [10]. GL was also provided by Sigma-Aldrich Pty. Ltd, Australia with the material specifications presented in Table 2.1.

Table 2.1 Physical properties of materials used in this study (based on material data sheets from Sigma-Aldrich Pty. Ltd, NSW, Australia)

Polymer	Molecular weight (g/mol)	Colour	Relative density (g/cm^3)	Other thermally related properties (°C)
PVA	89×10^3 to 98×10^3	Colourless	1.26	Melting point: 200
				Flash point: >113
ST	166×10^3	White	1.55	Gelatinisation point: 56–68
GL	92.09	Colourless	1.26	Boiling point: 182
				Flash point: 160

2.1.4 Halloysite Nanotubes (HNTs)

HNTs are natural nanoclays in kaolin family with hollow and tubular structures. HNTs have high aspect ratios and specific surface areas of 10–50 and 22.1–81.6 m^2/g, respectively [11–13]. Such nanofillers possess high thermal stability and moderate hydrophobicity, making them a good additive candidate to improve thermal and barrier properties [14, 15]. HNTs were donated by Imerys Ceramics Ltd, New Zealand, in the form of ultrafine particles with a relative density of 2.53 g/cm^3, which were used without any further purification.

2.1.5 Other Reagents

Magnesium bromide hexahydrate, magnesium nitrate hexahydrate, strontium chloride hexahydrate and barium chloride dehydrate salts were all purchased from Sigma-Aldrich Pty. Ltd, Australia. Saturated solutions of these salts were used to maintain the relative humidity (RH%) at a given level [16, 17] when water vapour transition rate (WVTR) and water vapour permeability (WVP) were evaluated. Chemical formulae and physical properties of these salts are summarised in Table 2.2.

Furthermore, ethanol solution with 100% concentration (Rowe Scientific Pty. Ltd, Australia), nitric acid with 70% concentration (Sigma-Aldrich Pty. Ltd, Australia) and glacial acetic acid with 96% concentration (Merck Pty. Ltd, Australia) were used as food simulants in migration tests according to European Commission Regulation (EU) No 10/2011 [18] and British Standard EN 1186-1 [19] to evaluate the overall migration rates and migration rates of nanofillers. For the same tests, standard solutions of aluminium and silicon ions were prepared from aluminium stock solution with the concentration of 1000 mg/L (Thermo Fisher Scientific Pty. Ltd, Australia) and silicon stock solution with the same concentration (High-Purity Standards, Inc.), respectively as control samples.

Table 2.2 Chemical formulae and physical properties of salts

Salt	Chemical formula	Molecular weight (g/mol)	Relative density (g/cm^3)	RH% at 25 °C
Magnesium bromide hexahydrate	$MgBr_2 \cdot 6H_2O$	292.20	2.00	30
Magnesium nitrate hexahydrate	$Mg(NO_3)_2 \cdot 6H_2O$	256.41	1.63	50
Strontium chloride hexahydrate	$SrCl_2 \cdot 6H_2O$	266.62	1.93	70
Barium chloride dehydrate	$BaCl_2 \cdot 2H_2O$	244.26	3.10	90

2.2 Manufacturing Process

Solution casting and melt blending are the most popular processing methods for PVA, PVA blends and PVA nanocomposites [20]. Moreover, solution casting process is particularly used for PVA/ST blends since 1980 because these polymers have melting temperatures close to their degradation temperatures, thus resulting in possible thermal degradation during the melt blending process [21, 22]. This phenomenon has been proven when comparing the number of published articles reporting the use of prevalent solution casting process to prepare PVA/ST blends over melting blending process according to Scopus data shown in Fig. 2.1. Consequently, solution casting process was used to manufacture neat PVA, PVA blends and PVA/ST/HNT bionanocomposite films in this study. The principle of solution casting process is very simple based on dissolving polymers in a suitable solvent-like water (e.g. hydrophilic polymers), and then mixing this solution with the suspension of well-dispersed nanofillers via continues mixing. Finally, the homogenous mixture is cast and dried at room temperature or in an oven to remove the solvent [23].

2.2.1 Polymer Blends

Neat PVA solution at the concentration of 5 wt%/v was prepared by dissolving 10 g PVA powder in 190 ml deionised water at 35 °C. This solution was heated gradually up to 85 °C for 3 h by continued stirring at 500 rpm with the aid of an IKA®-RCT magnetic stirrer. Equal amounts of clear solutions were poured in glass Petri dishes (diameter: 15 cm). Samples were dried for 48 h at 50 °C in an oven (see Fig. 2.2). The same procedure was followed to prepare plasticised PVA film (i.e. PVA/GL blends). GL solution (30 wt% content relative to the dry weight) was added to neat PVA solution by magnetic stirring during last 30 min of preparation process. Whereas, PVA/ST blend films were prepared by mixing 8 g PVA with 2 g ST as powders at room temperature, and the mixture was further dissolved in 190 ml deionised water accordingly with the same procedure mentioned earlier. PVA/ST blend films were plasticised with 30 wt% GL to produce PVA/ST/GL films by adding GL solution at last 30 min in the stirring process. Completely dried films were removed carefully from Petri dishes and kept in a desiccator with silica gel underneath them to prevent humidity absorption for at least a week before material characterisation.

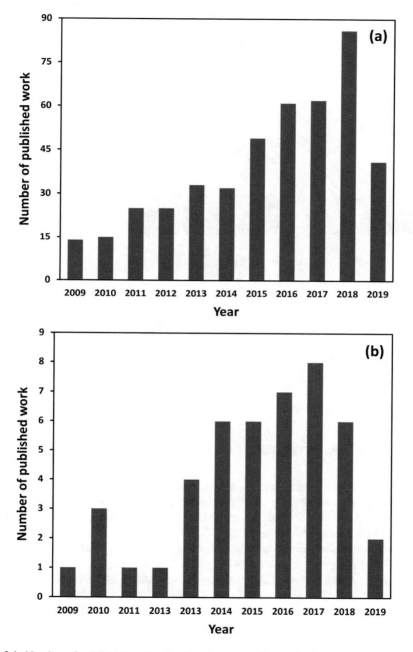

Fig. 2.1 Number of published work related to the preparation methods of PVA/ST blends: a solution casting process and b melt blending process according to Scopus data

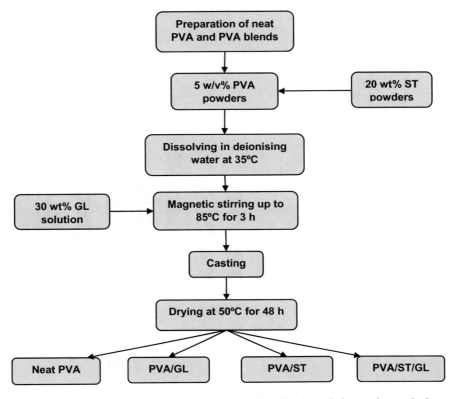

Fig. 2.2 Flow chart of preparing neat PVA and PVA blends using solution casting method

2.2.2 Bionanocomposites

PVA/ST/GL blends were adopted as the matrices for bionanocomposite films. HNT suspension was prepared by mixing weighted amounts of HNT powders (in range of 0.25, 0.50, 1, 3 and 5 wt%) in 100 ml deionised water using an IKA®RW20-mechanical mixer at 50 °C and 500 rpm for 2 h. Subsequently, the suspension was sonicated in an ultrasonicating bath ELMA Ti-H-5 model at 50 °C, with a 25-kHz frequency and a 90% power intensity for 1 h to get better HNT dispersion. Well-dispersed HNT suspension was added in a dropwise manner to 100 ml PVA/ST/GL blend solution via mechanical mixing at 500 rpm and 50 °C for 30 min. Such a solution was sonicated again at the same condition previously mentioned for additional 30 min in order to remove any air bubbles. Equal amounts of bionanocomposite solutions were poured into Petri dishes to dry them at 50 °C for 48 h, as shown in Fig. 2.3. All samples were further dried in a desiccator after removed from the Petri dishes.

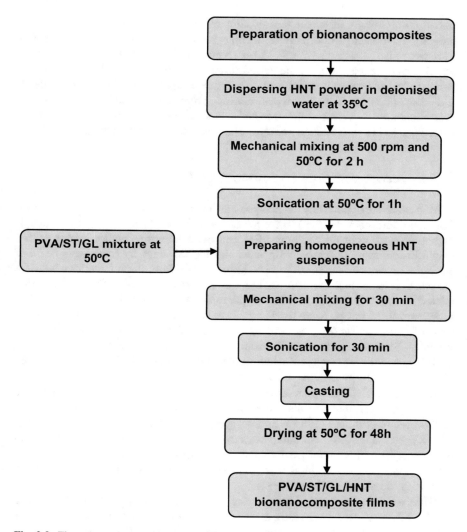

Fig. 2.3 Flow chart of preparing PVA/ST/GL/HNT bionanocomposites

2.3 Experimental Characterisation

2.3.1 Electron Microscopic Analysis

The holistic investigation of morphological structures of PVA/ST/GL blends and resulting bionanocomposite films was important to understand the compatibility between PVA, ST and GL in addition to the HNT dispersion within blend matrices. Consequently, scanning electron microscopy (SEM) was used to investigate the fracture morphology of tensile testing samples in order to evaluate their structure–property relationship. Furthermore, sample surfaces subjected to biodegradation tests were also examined with SEM observation before and after the tests. These samples were scanned during the period of the test progress as well at the first and third weeks, respectively. Moreover, the surface roughness of PVA/ST/GL blends and corresponding bionanocomposite films were investigated along with the determination of aspect ratios of both as-received HNT powders and those embedded HNTs in bionanocomposite films via atomic force microscopy (AFM).

2.3.1.1 Scanning Electron Microscopy (SEM)

SEM creates an opportunity to analyse and characterise heterogeneous materials from the microscale (10^{-6} m) to the nanoscale (10^{-9} m) to reveal structural details inaccessible by light microscopy. SEM can build a high-resolution image with a magnification ranging between 10 times and 10,000 times by focusing a beam of electrons on sample surfaces. The produced signals from the interactions between the electron beam and scanned surface can be interpreted to produce information related to crystallography, surface topography and material composition [24]. Moreover, SEM images were used to analyse the morphology of polymer blends in term of phase separation, as well as sizes and shapes of phase domains [25]. Whereas, SEM can be used for nanocomposites to characterise the fracture mode, as well as nanofillers shape, distribution, size and their chemical composition based on embedded energy dispersive spectroscopic (EDS) analysis [25, 26].

As-received HNT powders and fracture surfaces of neat PVA, PVA blends and PVA/ST/GL/HNT bionanocomposite films were investigated by a Tescan Mira 3 field emission scanning electron microscope (FE-SEM) at the accelerating voltage of 3 kV (see Fig. 2.4). All samples were coated with a layer of carbon (layer thickness: 10 nm) to improve the material contrast. Moreover, HNT dispersion within bionanocomposite films was evaluated by Oxford Instrument X-Max X-ray EDS with the aid of Aztec software.

NEON-40EsB field emission scanning electron microscope (FE-SEM), as seen in Fig. 2.5, was used to investigate the surface morphology of neat PVA, PVA blends and resulting bionanocomposite films at different HNT contents before, during and after biodegradation tests. FE-SEM was operated at an accelerating voltage of 2 kV after all samples were sputter coated with platinum layers (layer thickness: 3 mm) to improve the image contrast during scanning.

2.3.1.2 Atomic Force Microscopy (AFM)

AFM is one of the probe-based scanning methods that has been developed since 1980s [27]. In AFM, the static deflection of a cantilever to the dynamic excitation and detection of the cantilever-tip oscillation have been monitored with different modes either in vacuum, liquids or ambient conditions leading to many convincing

Fig. 2.4 Tescan Mira 3 field emission scanning electron microscope (FE-SEM)

results at an atomic level on nanoscale [28, 29]. These monitored signals can be used to image surface topography, as well as map changes in mechanical, electrical and magnetic properties of material surfaces [29]. As compared with other scanning techniques, AFM has a number of advantages such as 3D quantitative and qualitative analyses with non-destructive sample materials and imaging in different media and liquid dispersants [30].

Morphological structures and surface roughness of PVA/ST/GL blends and their bionanocomposite films at different HNT contents were also examined via a Bruker Dimension Fastscan AFM system depicted in Fig. 2.6. A drop of sonicated suspension of as-received HNT powders was deposited on the mica substrate for analysis. On the other hand, PVA/ST/GL blends and their bionanocomposite films were glued on glass substrates with a carbon tape. A tapping mode was used for AFM measurements with a TESPA probe at the scanning rate of 2 Hz, as well as a nominal resonant frequency of 525 kHz with images being captured in an ambient condition. Surface roughness of bionanocomposite films and HNT aspect ratios for as-received powders and those embedded within bionanocomposite films were individually determined with the aid of NanoScope Analysis 1.90 software.

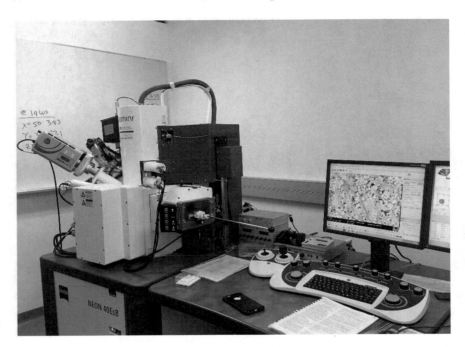

Fig. 2.5 NEON-40EsB field emission scanning electron microscope (FE-SEM)

2.3.2 Mechanical Properties

Tensile tests are used mostly to characterise mechanical properties of material such as tensile strength, Young's modulus and elongation at break. Generally, mechanical properties can vary with material composition, sample thickness, sample preparation method, load cell used, grip types and strain rate [31]. Generally, the reinforcing effect of microfillers or nanofillers into neat polymers can significantly alter these properties by enhancing mechanical strength and toughness, as well as restricting the elongation rate [32]. Consequently, mechanical properties of composite materials are correlated with effective interfacial bonding between matrices and fillers for load transfer, filler loading, aspect ratio, dispersion and orientation within composites relative to applied load direction [33].

In this study, Lloyd-LR10K universal testing machine was employed to assess mechanical properties of neat PVA, PVA blends and resulting bionanocomposite films at different HNT contents according to ASTM D882 standard, as shown in Fig. 2.7. Rectangular samples for all material films were cut in size of 100 mm length, 20 mm width and 100–120 μm thickness. For each material composition, 5–8 samples were tested to get average data with reported standard deviations. Crosshead speed and gauge length of samples were set to 10 mm/min and 50 mm, respectively.

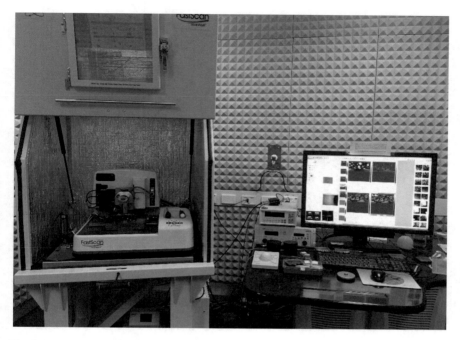

Fig. 2.6 Bruker dimension Fastscan AFM system

2.3.3   Thermal Properties

Thermal properties of any material can be determined via differential scanning calorimetry (DSC) and thermogravimetric analysis (TGA) [34]. DSC is a technique to measure the heat flow of materials as a function of time or temperature. DSC results can be used to specify different material properties such as melting temperature (T_m), which is a transformational point from a solid to liquid state, glass transition temperature (T_g) representing the transformation of amorphous materials during heating (soft) and cooling (brittle) cycle and crystallinity rate (X_c), known as the crystallinity rate of materials [34, 35]. Whereas, TGA is a quantitative analysis of materials based on monitoring the mass of samples during heating or cooling in a predefined atmosphere to study the reactions of materials decomposition [34, 36].

Thermal properties of neat materials (i.e. PVA, ST and HNTs), PVA blends and resulting bionanocomposites at different HNT contents were determined via TGA and DSC in the TGA/DSC 1 STAR^e System, as illustrated in Fig. 2.8. The weighed samples of 5–15 mg were heated at the heating rate of 10 °C/min from 35 to 600 °C in argon atmosphere (flow rate: 30 ml/min). The decomposition temperatures of materials were determined from TGA curves. These temperatures were detected at the weight losses of 5, 50 and 90%, which were referred to as $T_{5\%}$, $T_{50\%}$ and $T_{90\%}$, respectively. Whereas, the degradation temperatures were determined from

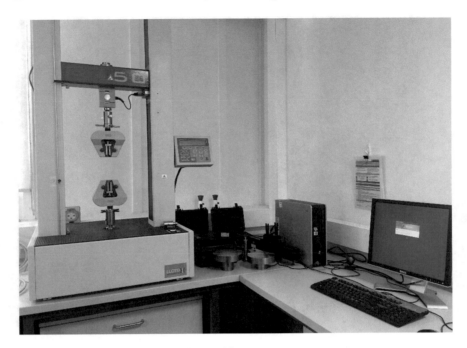

Fig. 2.7 Lloyd-LR10K universal testing machine

derivative thermogravimetric (DTG) curves, and represented by T_{d1}, T_{d2}, T_{d3} and T_{d4}. Furthermore, T_g, T_m, crystallisation temperature (T_c) and X_c were obtained from DSC thermograms during the heating scans. X_c can be calculated according to the equation given below [37]:

$$X_c = \frac{\Delta H_m - \Delta H_c}{\Delta H_m^\circ} \times \frac{100}{1 - w_f} \tag{2.1}$$

where ΔH_m and ΔH_c are the melting and crystallisation enthalpies of polymer matrices in bionanocomposites, respectively. w_f is the weight fraction of nanofillers. In particular, ΔH_m° is the melting enthalpy of 100% crystalline PVA (i.e. $\Delta H_m^\circ = 142\,\text{J/g}$ [38]).

2.3.4 X-Ray Diffraction (XRD) Analysis

XRD analysis is one of non-destructive characterisation techniques with high reliability, simplicity and quantitative information, particularly used in nanocomposite fields in term of polymer–filler interactions [39]. The presence of nanofillers

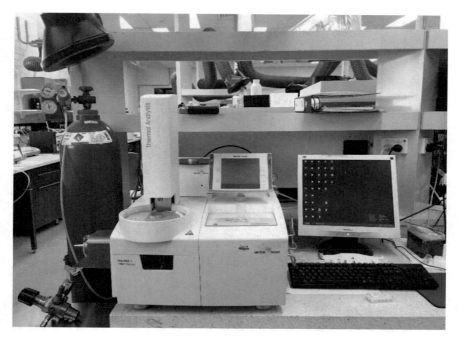

Fig. 2.8 TGA/DSC 1 STAR^e system for thermal analysis

within polymer matrices, generally acting as nucleating agents, leads to the improvement of nanocomposite crystallinity by changing interfacial interactions according to XRD spectra [23, 39]. Moreover, XRD analysis is successfully used to characterise crystalline and amorphous structures in neat polymers and polymer blends. Different scattering angles can be set up for XRD operation at wide, small and ultra-small angles to provide structural information in three different scales of 10, 100 and 1000 nm, respectively [23, 39].

The types of nanocomposite structures (i.e. intercalated and/or exfoliated) were investigated using XRD analysis. XRD patterns of as-received HNT powders and PVA/ST/GL/HNT bionanocomposite films at different HNT contents were investigated via a Bruker D8 advance diffractometer (Cu-Kα source, wave length $\lambda = 0.1541$ nm), as depicted in Fig. 2.9, at an accelerating voltage and current of 40 kV and 40 mA, respectively. The material samples were scanned in a diffraction angle (2θ) range of 5°–40° at the scan rate of 0.013°/s. Bragg's law was used to calculate interlayer spacing d for both as-received HNT powders and embedded HNTs in bionanocomposite films at different HNT contents according to the following equation [40, 41]:

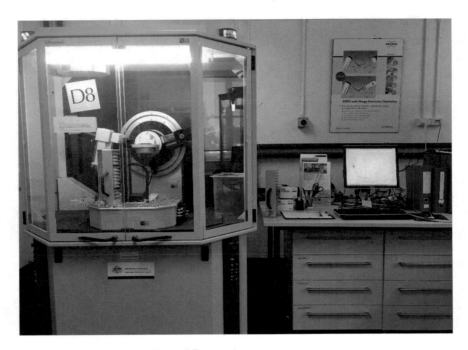

Fig. 2.9 Bruker D8 advance X-ray diffractometer

$$n\lambda = 2d \sin \theta \qquad (2.2)$$

where n is an integer used for incident X-ray beam.

2.3.5 Fourier Transformation Infrared (FTIR) Analysis

FTIR is widely used to understand the interaction bonding between polymer blend and nanocomposite constituents by recording absorption bands corresponding to fundamental stretching and bending vibrations of their structural units [42, 43]. FTIR has been widely used for analysing materials structure because small samples required for the tests without preparation, different sample forms can be used like powders, solids and liquids by means of fast and easy recording the spectra as a relatively inexpensive technique [42].

FTIR was used to investigate molecular bonding interactions for neat materials (PVA, ST, GL and HNTs), PVA blends and PVA/ST/GL/HNT bionanocomposite films at different HNT contents. All samples were tested via a PerkinElmer 100 FTIR spectrometer mounted with the attenuated total reflection (ATR) accessory, which enables samples to be examined directly in different states without further

Fig. 2.10 PerkinElmer 100 FTIR spectrometer

preparation in FTIR at the scanning resolution of 4/cm and a wavenumber range of 4000–500/cm, as illustrated in Fig. 2.10.

2.3.6 UV-Vis Spectra

Optical properties of polymer composites can be evaluated by different ways like refractive index, UV absorption, transparency and luminescence based on their applications such as biomedical and packaging applications, as well as colour filters and light emitting diodes [44]. Most polymer composites with microfillers are opaque because of light scattering taking place for similar fillers due to the reduction in light transmittance [45]. For polymer nanocomposites, the effect of nanofillers may diminish because the dimensions of nanofillers appear to be mostly lower than the light wavelength resulting in better optical properties [45]. Moreover, optical properties of nanocomposites are governed by the size and distribution of nanofillers within matrices, as well as absorption properties of nanofillers [46].

In food packaging applications, the transparency level of packaging materials is vital. The light transmittance of neat PVA, PVA blends and PVA/ST/GL/HNT bionanocomposite films at different HNT contents was determined with the aid of an ultraviolet-visible (UV-vis) spectrometer (Jasco-V670), as depicted in Fig. 2.11. Light transmittance ($T\%$) was measured with a wavelength range of 200–800 nm to cover ultraviolet (200–400 nm), visible (400–700 nm) and infrared (700–800 nm) wavelengths [45, 47] at a scan rate of 200 nm/min. In addition, a glass plate was utilised as a reference sample. Three samples were investigated for each material batch for data repeatability. Moreover, additional visual comparison in terms of transparency was made by using captured digital images of Curtin University logo.

2.3.7   Barrier Properties

Barrier properties can be defined as physical resistance of a material against the passage of permeable molecules through the material [32]. These properties are one of the most important requirements for packed materials where they should possess good barrier properties to protect packaged products and extend their shelf life [48, 49]. Generally, the permeation process of small molecules is governed by the solubility, diffusivity and morphology of materials. Moreover, the process mostly happens through the amorphous phase of materials by sorption and diffusion [48]. The incorporation of well-dispersed nanofillers within polymer matrices may generate tortuous paths, which can produce a barrier structure against permeable

Fig. 2.11 Ultraviolet-visible (UV-vis) spectrometer

molecules such as water vapour and gases with the variation of solubility, diffusivity and morphological structures of polymer matrices. The high tortuosity means high barrier properties and lower permeability [48, 49]. Consequently, WVTR and WVP of neat PVA, PVA blends and PVA/ST/GL/HNT bionanocomposite films at different HNT contents were evaluated at a various range of temperatures and RH gradients. Furthermore, the gas permeabilities of these aforementioned films were also examined against oxygen and air.

2.3.7.1 Water Vapour Transmission and Water Vapour Permeability

According to ASTM E96M-16 standard, WVTR and WVP were determined for neat PVA, PVA blends and PVA/ST/GL/HNT bionanocomposite films at different HNT contents. Circular samples of these pre-dried films were cut and sealed on the mouth of laboratory bottle of borosilicate glass filled with distilled water (100% RH). The gap between the surfaces of material samples and water in each bottle should be at least 20 mm to avoid any direct contact. These bottles were kept in an air-circulating oven at 25 ± 1 °C and 50% $\pm$ 2% RH for two weeks. The weight loss of bottle was recorded daily to establish the relationship between weight loss and time for determining the slope of the steady-state period in order to calculate the WVTR according to the equation below:

$$\text{WVTR} = \frac{G/t}{A} \tag{2.3}$$

where G is the weight loss of bottle (g), t is the time (h), G/t is the slope of linear regression (g/h) and A is the testing sample area (m^2). These data were further employed to calculate the WVP as follows:

$$\text{WVP} = \frac{\text{WVTR} \times l}{S \times \Delta\text{RH}} \tag{2.4}$$

where l is the sample thickness (m) determined as the average thickness value in different positions when randomly selected on the sample surfaces, S is the saturated water vapour pressure (kPa) and ΔRH is the relative humidity gradient across the sample (fixed at 50%), which is the difference of relative humidity between those inside and outside permeability bottles. All associated tests were conducted three times to obtain average data of WVTR and WVP with associated standard deviations.

The temperature effect on WVTR and WVP of neat PVA, PVA blends and bionanocomposite films at different HNT contents was examined by running the tests at different temperatures of 25, 35, 45 and 55 °C and a constant RH level of 50% $\pm$ 2% according to the procedure mentioned previously. The RH gradient was maintained at a constant level of 50% $\pm$ 2% by using the saturated salt solution of $Mg(NO_3)_2 \cdot 6H_2O$ [16, 17] out of the test bottle in order to keep the RH level at

50%, as opposed to the RH level of 100% inside the bottle due to being filled with distilled water. The saturated salt solution was prepared by dissolving a sufficient amount of salt in distilled water at a boiling point, and the solution was gradually cooled with additional salt for initialising the precipitation. This solution was kept to stabilise for 1–2 weeks before further use [50].

The effect of RH level on the WVTR and WVP of neat PVA, PVA blends and PVA/ST/GL/HNT bionanocomposite films at different HNT contents was also investigated with RH gradients of 70, 50, 30 and 10% ± 2% at a given temperature of 25°C. The RH level remained at 100% inside the test bottle by filling it with distilled water. On the other hand, various RH levels were generated out of the test bottle by using saturated salt solutions, as described in earlier work [16, 17]. The details of the salts used as well as RH% and ΔRH% are summarised in Table 2.3:

2.3.7.2 Gas Permeability

The dynamic method was used to assess gas permeabilities of neat PVA, PVA blends and PVA/ST/GL/HNT bionanocomposite films at different HNT contents. This method depends on the gas diffusion through material films [51, 52]. The preconditioned samples at 25°C with the RH level of 50% were tightly fixed between two Teflon rings of permeability cell (Thermo Fisher Scientific Pty. Ltd, Australia). Gas permeability was determined for both oxygen and air. The test gas was continuously circulated in one chamber of permeability cell (channels A and B) at a controlled flow rate to keep a constant pressure gradient at 80 kPa. Nitrogen carrier gas was used at the same flow rate through channel C of another permeability cell chamber. The channel D of the same chamber was used to collect the gas mixture (i.e. test gas and carrier gas) and injected directly to the gas chromatograph (GC), as shown in Fig. 2.12. The GC (Shimadzu GC-2014 with 5 Å molecular sieve column) was equipped with thermal conductivity detector and operated at a column temperature of 50°C and a helium flow rate of 25 ml/min. The test gas concentration in the mixture was calculated as a percentage of detected peaks. Subsequently, the gas permeability can be determined according to the following equation [51, 52]:

$$\text{Gas permeability} = \frac{a \times x \times V}{S \times t \times \Delta P} \qquad (2.5)$$

Table 2.3 Saturated salt solutions with various RH% levels

Salt	RH% at 25 °C	ΔRH% for permeability tests
$MgBr_2 \cdot 6H_2O$	30	70
$Mg(NO_3)_2 \cdot 6H_2O$	50	50
$SrCl_2 \cdot 6H_2O$	70	30
$BaCl_2 \cdot 2H_2O$	90	10

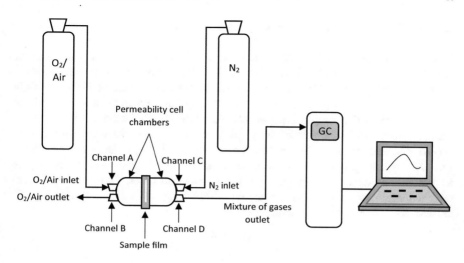

Fig. 2.12 Gas permeability apparatus

where *a* is the concentration percentage of test gas, *x* is the film thickness (m), *V* is the volume of permeability cell (m^3), *S* is the surface area of test films (m^2), *t* is the time of gas circulation in permeability cell (approximately 60 min) and ΔP is the gas pressure gradient (kPa).

2.3.8 *Water Resistance*

Generally, most biodegradable polymers are water-sensitive polymers due to the presence of a good number of hydrophilic groups, which may interact with water molecules particularly in water-soluble polymers [53]. Moreover, the absorbed water induces the alteration of dimensions and properties like mechanical strength and thermal stability due to the plasticisation effect of water [54]. Consequently, it is important to develop water resistance for biodegradable polymers to keep their properties stable during applications [54] by coating [55], using cross-linking agents [56] and reinforcing with nanofillers [57]. For example, when nanocomposite materials are developed for food packaging applications, their interaction with water should be at a minimum level to improve the shelf life of packed products. As such, water absorption capacity, water solubility and water contact angle were considered as key water-related properties for the evaluation.

2.3.8.1 Water Absorption Capacity (W_a)

Water absorption capacity (W_a), which is also known as water uptake, can be determined based on ASTM D570-98 standard. Square samples of neat PVA, PVA blends and PVA/ST/GL/HNT bionanocomposite films at different HNT contents in size of 2×2 cm^2 were pre-dried to remove any residual moisture at 50 °C for 24 h. The initial dry weight (W_o) of samples was recorded after cooled to room temperature in a desiccator filled with silica gel. These samples were immersed in 100 ml distilled water for 24 h to reach an equilibrium state at ambient conditions. The wetting samples were removed from water and dried gently with tissue papers to remove any excessive amount of water on their surfaces. These samples were weighed again to record their weight with absorbing water (W_t). W_a can be calculated according to the following equation:

$$W_a(\%) = \frac{W_t - W_o}{W_o} \times 100\% \qquad (2.6)$$

Three samples were tested for each material batch in order to obtain the average data with calculated standard deviations.

2.3.8.2 Water Solubility (W_s)

The wetting samples from water absorption tests were used to measure the water solubility (W_s) according to ASTM D570-98 standard. Such samples were dried in a vacuum oven at 60 °C for 24 h to evaporate absorbed water. The dried samples were measured again to record the dry weight (W_d) after cooled to room temperature in a desiccator filled with silica gels as well. W_s can be further calculated as follows:

$$W_s(\%) = \frac{W_t - W_d}{W_t} \times 100\% \qquad (2.7)$$

Average values with calculated standard deviations were recorded based on three repeated tests for each material batch.

2.3.8.3 Water Contact Angle

The wettability of materials can be generally characterised by water contact angle [58]. A Tensiometer KSV-CAM 101 was employed to measure water contact angles of neat PVA, PVA blends and PVA/ST/GL/HNT bionanocomposite films at different HNT contents (see Fig. 2.13). The droplet of 2 µl deionised water was dropped on the film surface by using a Sessile Drop Half-Angle^{TM} tangent line

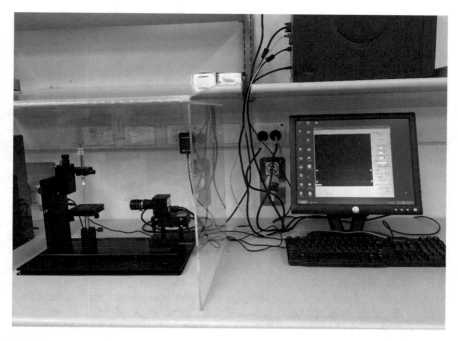

Fig. 2.13 Tensiometer KSV-CAM 101 for measurements of water contact angle

method [59, 60] to determine the hydrophilicity of film materials. Average values in relation to water contact angles were reported based on five droplets at randomly selected positions on different material films.

2.3.9   Migration Tests

When a packaging material gets in contact with foodstuffs, the migration of substances from packaging materials should be taken into account based on the criterion of an acceptable threshold of no more than 60 mg/kg of foodstuffs [18]. The migrated constituents could be a wide range of substances such as adhesives, coatings, solvents, surfactants and plasticisers, as well as microfillers and nanofillers [18]. Consequently, the overall migration rate of PVA/ST/GL blends and their bionanocomposite films, as well as the migration rate of HNTs from bio-nanocomposite films were studied together in this study.

European Commission Regulation (EU) No 10/2011 [18] was followed to evaluate the migration rates of PVA/ST/GL blends and their bionanocomposite films at different HNT contents. Three food simulants were selected to mimic hydrophilic, acidic and lipophilic foodstuff conditions, namely 10% (v/v) ethanol solution (simulant A), 3% (w/v) acetic solution (simulant B) and 50% (v/v) ethanol solution (simulant D1), respectively, based on European Commission Regulation (EU) No 10/2011 [18]. According to British Standard EN 1186-1 [19], the material samples of PVA/ST/GL blends and their bionanocomposite films in size of 1 square decimetre (dm^2) were completely immersed in glass bottles filled with 100 ml of each food simulants. Three samples of each material batch and food simulants were prepared and kept in an air-circulating oven at 40 °C over the period of 10 days. After 10 days, the bottles containing samples were removed from the oven, and all materials were cooled down to room temperature at ambient conditions before opening their covers for minimising the evaporation of food simulants. Samples were removed gently from bottles, and food simulants were evaporated, and their residual traces were dried completely overnight at 105 °C. The residues were cooled to room temperature, and then weighed with an analytical balance (± 0.0001 g precision) to calculate the overall migration rate in comparison with the overall migration limit (OML) of 60 mg/kg [18].

A similar procedure of overall migration rate was also performed to study the migration rate of HNTs. The dried residues of food simulants were analysed using inductivity coupled plasma-optical emission spectroscopy (ICP-OES), as illustrated in Fig. 2.14, to identify the presence of aluminium ions (Al^+) and silicon ions (Si^+), as an indicator of the HNT migration from bionanocomposite films. At least three residue samples were utilised with ICP-OES after digested with 15 ml HNO_3 solution at the concentration of 3% for 2.5 h at 95 °C to convert Al and Si elements into an ionic state, which further diminished the effect of blend matrices and any other contaminations. PerkinElmer-Optima 8300 ICP-OES was employed for the corresponding analysis along with the operation parameters being summarised in Table 2.4. Two blank solutions were used for the calibration purpose consisting of distilled water and 3% HNO_3 solution. Furthermore, three controlled samples were prepared from a serial dilution of standard stock solution with Al^+ and Si^+ for the comparison and calibration. 1000 mg/L concentrated solutions containing both Al^+ and Si^+ were diluted gradually to prepare Al^+ and Si^+ control samples at the concentrations of 10, 5 and 1 mg/L, respectively.

2.3.10 Soil Burial Tests

Biodegradation process can be defined simply as a process of material decomposition into a simple compound by the microorganisms in suitable circumstance of temperature, humidity and pH level [61]. Moreover, this process depends on the properties of materials such as chemical composition, molecular weight, crystallinity, etc. [62]. Biodegradation rate can be determined by several ways like weight loss, visual changes, variation of material properties and formation of final products like CO_2 [62, 63].

The biodegradability of material films was evaluated by means of soil burial degradation tests according to the procedure developed by Thakore et al. [64]. Square samples (3 × 3 cm^2) of neat PVA, PVA blends and PVA/ST/GL/HNT

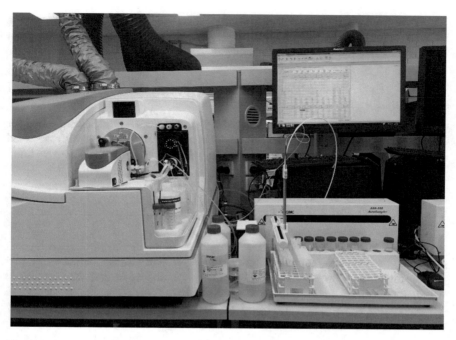

Fig. 2.14 Inductivity coupled plasma-optical emission spectroscopic unit

Parameter	Value
Plasma viewing mode	Radial and Axial
Plasma gas flow rate (L/min)	15.00
Auxiliary gas flow rate (L/min)	0.50
Nebuliser gas flow rate (L/min)	0.60
Pump flow rate (mL/min)	1.50
Radiofrequency (Watts)	1400

Table 2.4 Operation parameters of ICP-OES for HNT migration tests

bionanocomposite films at different HNT contents were weighed with aid of an analytical balance to record their initial weights (W_o). These samples were buried at 5 cm under the surfaces in plastic containers in 2 L capacity after filled with sieved agricultural soil purchased from a local plant nursery. These containers were kept at room temperature with a RH level of 40–50% and using a humidity meter to monitor the RH at the same level by sprinkling water to reduce RH if necessary. The biodegradability results were recorded as a weight loss with time over a period of six months (i.e. 24 weeks). In initial three months, the samples were removed from the containers on a weekly basis, and gently washed with distilled water to remove soil from their surfaces. This routine was extended to be once every three weeks in the following three months. The clean samples were completely dried at

70 °C for 24 h to evaporate any moisture absorbed from the soil and/or in a washing process. The dried samples were weighed again to record the dry weight (W_d). Biodegradation rate can be calculated according to the following equation:

$$\text{Biodegredation rate } (\%) = \frac{W_o - W_d}{W_o} \times 100\% \qquad (2.8)$$

The weights of samples were measured in (g), and three samples were tested for each material composition for recording their average data along with associated standard deviations along with the additional samples used for SEM observation.

2.3.11 Food Packaging Tests

Food packaging materials should have specific requirements like good mechanical properties, acceptable thermal stability with low cost. These materials should possess high barrier properties as well to prevent gas and water molecules to diffuse within the package and contact the products [61, 65]. These requirements are already available for petroleum-based polymers when considering one of major reasons behind the plastic waste problem. Nowadays, developing food packaging materials based on biodegradable polymers has been so attractive to cut off the solid plastic wastes [66]. The best scenario to follow for making biodegradable polymers meet the food packaging requirements is that the use of bionanocomposites should improve their weak mechanical, thermal and barrier properties [3]. In this work, the developed bionanocomposites was holistically investigated as a potential food packaging material.

Neat PVA, PVA/ST/GL blend and PVA/ST/GL/1 wt% HNT bionanocomposite films were used for food packaging tests. These films were manufactured in size of 23.0 × 24.0 cm according to the same procedure mentioned earlier in Sect. 2.2. These films were double sealed (i.e. two sealing lines) with a hot sealing machine Pro-Line-C1 to avoid any gas leaking. Avocados with a lipid content of 20% (≈18.7/100 g) [67] and peaches with pH level ≤ 3.5 [68] were purchased from local markets with a similar size, colour and appearance, as well as defect-free to minimise the differences between samples during the tests, as shown in Fig. 2.15. Packaged fruits of peaches and freshly cut avocados were stored in a fridge at 8 °C with a RH level of 85% over the period of 14 days [67, 68], which was subjected to the data recording of daily weight loss as an indicator for the shelf life of fruits [69].

Fig. 2.15 Flow chart of the food packaging test procedure

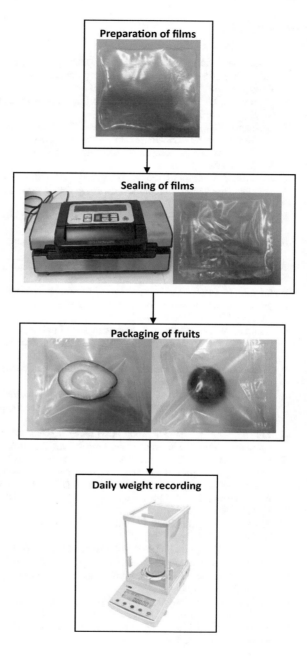

References

1. Sustainable Packaging Coalition (2011) Definition-of-sustainable-packaging. Available online at: https://sustainablepackaging.org/wp-content/uploads/2017/09/Definition-of-Sustainable-Packaging.pdf. Accessed 25 March 2020
2. Magnier L, Crié D (2015) Communicating packaging eco-friendliness. Int J Retail Distr Manag 43(4/5):350–366
3. Cerqueira MA, Vicente AA, Pastrana LM (2018) Nanotechnology in food packaging: opportunities and challenges. In: Cerqueira MAPR, Castro LMP, Vicente AAMOS (eds) Nanomaterials for food packaging: materials, processing technologies, and safety issues. Elesvier, pp 1–11
4. Sapalidis AA, Katsaros FK, Kanellopoulos NK (2011) PVA/montmorillonite nanocomposites: development and properties. In: Cuppoletti J (ed) Nanocomposites and polymers with analytical methods. InTechOpen, China
5. Chiellini E, Corti A, D'Antone S, Solaro R (2003) Biodegradation of poly (vinyl alcohol) based materials. Prog Polym Sci 28:963–1014
6. Avérous L (2004) Biodegradable multiphase systems based on plasticized starch: a review. J Macromol Sci Part C 44(3):231–274
7. Corre DL, Bras J, Dufresne A (2010) Starch nanoparticles: a review. Biomacromol 11:1139–1153
8. Schmitt H, Prashantha K, Soulestin J, Lacrampe MF, Krawczak P (2012) Preparation and properties of novel melt-blended halloysite nanotubes/wheat starch nanocomposites. Carbohyd Polym 89(3):920–927
9. Rahman WAWA, Sin LT, Rahmat AR, Samad AA (2010) Thermal behaviour and interactions of cassava starch filled with glycerol plasticized polyvinyl alcohol blends. Carbohyd Polym 81:805–810
10. Siddaramaiah BR, Somashekar R (2003) Structure-property relation in polyvinyl alcohol/starch composites. J Appl Polym Sci 91:630–635
11. Idumah CI, Hassan A, Ogbu J, Ndem JU, Nwuzor IC (2019) Recently emerging advancements in halloysite nanotubes polymer nanocomposites. Compos Interf 26(9):751–824
12. Liu M, Jia Z, Jia D, Zhou C (2014) Recent advance in research on halloysite nanotubes-polymer nanocomposite. Prog Polym Sci 39(8):1498–1525
13. Mousa MH, Dong Y, Davies IJ (2016) Recent advances in bionanocomposites: preparation, properties, and applications. Int J Polym Mater Polym Biomater 65(5):225–254
14. Xie Y, Chang PR, Wang S, Yu J, Ma X (2011) Preparation and properties of halloysite nanotubes/plasticized Dioscorea opposita Thunb. starch composites. Carbohyd Polym 83:186–191
15. Zhang Y, Tang A, Yang H, Ouyang J (2016) Applications and interfaces of halloysite nanocomposites. Appl Clay Sci 119:8–17
16. Wexler A, Hasegawa S (1954) Relative humidity-temperature relationships of some saturated solutions in the temperature range 0° to 50 °C. J Res Nation Bureau Stand 53(1):19–26
17. Young JF (1967) Humidity control in the laboratory using salt solutions—a review. J Appl Chemist 17:241–245
18. European Union Commission regulation (EU) No 10/2011 (2011) On plastic materials and articles intended to come into contact with food. Offic J Europ Union L12:1–89
19. British Standard 1186-1(2002) Materials and articles in contact with foodstuffs-plastics. Part 1: guide to the selection of conditions and test methods for overall migration. Europ Commit Standard 1–52
20. Zhou J, Ma Y, Ren L, Tong J, Liu Z, Xie L (2009) Preparation and characterization of surface crosslinked TPS/PVA blend films. Carbohyd Polym 76(4):632–638

21. Jayasekara R, Harding I, Bowater I, Christie GBY, Lonergan GT (2004) Preparation, surface modification and characterisation of solution cast starch PVA blended films. Polym Testing 23(1):17–27

22. Sin LT, Rahman WAWA, Rahmat AR, Khan MI (2010) Detection of synergistic interactions of polyvinyl alcohol–cassava starch blends through DSC. Carbohyd Polym 79(1):224–226

23. Wang L, Li J, Hong R, Li H (2013) Synthesis, surface modification, and characterization of nanoparticles. In: Thomas S, Malhotra SK, Goda K, Sreekala MS (eds) Polymer composites. Wiley, Germany, pp 13–51

24. Goldstein JI, Newbury DE, Echlin P, Joy DC, Lyman CE, Lifshin E, Sawyer L, Michael JR (2003) Scanning electron microscopy and x-ray microanalysis, 3rd edn. Springer Science, New York

25. Kundu S, P Jana P, De D, Roy M (2015) SEM image processing of polymer nanocomposites to estimate filler content. Paper presented at IEEE international conference on electrical, computer and communication technologies (ICECCT), IEEE, Coimbatore, India, 5–7 Mar 2015

26. Dikin DA, Kohlhaas KM, Dommett GHB, Stankovich S, Ruoff RS (2006) Scanning electron microscopy methods for analysis of polymer nanocomposites. Microsc Microanal 12(Supp 2):674–675

27. Schwarz UD (2012) Noncontact atomic force microscopy. Beilst J Nanotechnol 3:172–173

28. Lozano JR, Garcia R (2008) Theory of multifrequency atomic force microscopy. Phys Rev Letter 100(7):076102

29. Paulo ÁS, García R (2002) Unifying theory of tapping-mode atomic-force microscopy. Phys Rev B 66(4):041406

30. Bandyopadhyay S, Samudrala SK, Bhowmick AK, Gupta SK (2008) Applications of atomic force microscope (AFM). In: Seal S (ed) Field of Nanomaterials and nanocomposites, in functional nanostructures: processing, characterization, and applications. Springer, New York, pp 504–568

31. ASTM, Standard test method for tensile properties of thin plastic sheeting. 2012. American National Standards Institute (ANSI), New York

32. Zeng Q, Yu A (2013) Theory and Simulation in Nanocomposites, In: Thomas SK, Malhotra SK, Goda K, Sreekala MS (eds) Polymer composites. Wiley, Germany, pp 53–74

33. Nielsen LE, Landel RF (1994) Mechanical properties of polymers and composites, 2nd edn. Taylor & Francis, New York

34. Corcione CE, Greco A, Frigione M, Maffezzoli A (2013) Polymer nanocomposites characterized by thermal analysis techniques. In: Thomas S, Malhotra SK, Goda K, Sreekala MS (eds) Polymer composites. Wiley, Germany, pp 201–217

35. Hammer A (2013) Thermal analysis of polymers. Mettler Toledo, Greifensee

36. Mena ME, Estrada RF, Torres LA, Tinajero R, Palafox L (2018) Characterization of polymer nanocomposites with thermal analysis and spectrum techniques. Adv Mater 12–15

37. Dong Y, Marshall J, Haroosh HJ, Mohammadzadehmoghadam S, Liu D, Qi X, Lau KT (2015) Polylactic acid (PLA)/halloysite nanotube (HNT) composite mats: influence of HNT content and modification. Compos A App Sci Manuf 76:28–36

38. Sreekumar PA, Al-Harthi MA, De SK (2012) Effect of glycerol on thermal and mechanical properties of polyvinyl alcohol/starch blends. J Appl Polym Sci 123:135–142

39. Venkateshaiah A, Nutenki R, Kattimuttathu S (2016) X-ray diffraction spectroscopy of polymer nanocomposites. In: Thomas S, Ponnamma D (eds) Spectroscopy of polymer nanocomposites. Elsevier, UK, pp 410–451

40. Zagho MM, Khader MM (2016) The impact of clay loading on the relative intercalation of poly(vinyl alcohol)-clay composites. J Mater Sci Chem Eng 4:20–31

41. Dong Y, Bickford T, Haroosh HJ, Lau KT, Takagi H (2013) Multi-response analysis in the material characterisation of electrospun poly (lactic acid)/halloysite nanotube composite fibres based on Taguchi design of experiments: fibre diameter, non-intercalation and nucleation effects. Appl Phys A 112:747–757

42. Jaleh B, Fakhri P (2016) Infrared and Fourier transform infrared spectroscopy for nanofillers and their nanocomposites. In: Thomas S, Ponnamma D (eds) Spectroscopy of polymer nanocomposites. Elsevier, UK, pp 112–129
43. Bokobza L (2017) Spectroscopic techniques for the characterization of polymer nanocomposites: a review. Polymers 10(1):7
44. Venkatachalam S (2016) Ultraviolet and visible spectroscopy studies of nanofillers and their polymer nanocomposites. In: Thomas S, Ponnamma D (eds) Spectroscopy of polymer nanocomposites. Elsevier, UK, pp 130–157
45. Paiva LBD, Morales AR (2013) Characterization of nanocomposites by optical analysis. In: Thomas S, Malhotra SK, Goda K, Sreekala MS (eds) Polymer composites. Wiley, Germany, pp 147–161
46. Srivastava S, Haridus M, Basu JK (2008) Optical properties of polymer nanocomposites. Bull Mater Sci. 31(3):213–217
47. ASTM, Standard practice for fluorescent ultraviolet (UV) lamp apparatus exposure of plastics. 2013. American National Standards Institute (ANSI), New York
48. Saritha A, Joseph K (2013) Barrier properties of nanocomposites. In: Thomas S, Malhotra SK, Goda K, Sreekala MS (eds) Polymer composites. Wiley, Germany, pp 185-200
49. Feldman D (2013) Polymer nanocomposite barriers. J Macromol Sci A 50(4):441–448
50. Winston PW, Bates D (1960) Saturated solutions for the control of humidity in biological research. Ecology 41(1):232–237
51. Khwaldia K, Sylvie Banon S, Desobry S, Hardy J (2004) Mechanical and barrier properties of sodium caseinate–anhydrous milk fat edible films. Int J Food Sci Technol 39:403–411
52. Imran M, Revol-Junelles AM, René N, Jamshidian M, Akhtar MJ, Arab-Tehrany E, Jacquot M, Desobry S (2012) Microstructure and physico-chemical evaluation of nano-emulsion-based antimicrobial peptides embedded in bioactive packaging film. Food Hydrocolloid 29:407–419
53. Enderby JA (1954) Water absorption by polymers. Trans Faraday Soc 51:106–116
54. Baschek G, Hartwig G, Zahradnik F (1999) Effect of water absorption in polymers at low and high temperatures. Polymer 40:3433–3441
55. Khwaldia K, Arab-Tehrany E, Desobry S (2010) Biopolymer coatings on paper packaging materials. Compreh Rev Food Sci Food Safet 9(1):82–91
56. Maiti S, Ray D, Mitra D (2012) Role of crosslinker on the biodegradation behavior of starch/polyvinylalcohol blend films. J Polym Environ 20(3):749–759
57. Abdollahi M, Alboofetileh M, Behrooz R, Rezaei M, Miraki R (2013) Reducing water sensitivity of alginate bio-nanocomposite film using cellulose nanoparticles. Int J Biol Macromol 54:166–173
58. Yuan Y, Lee TR (2013) Contact angle and wetting properties. In: Bracco G, Holst B (eds) Surface science techniques. Springer, Berlin, pp 3–34
59. Alipoormazandarani N, Ghazihoseini S, Nafchi AM (2015) Preparation and characterization of novel bionanocomposite based on soluble soybean polysaccharide and halloysite nanoclay. Carbohyd Polym 134:745–751
60. Sadegh-Hassani F, Nafchi AM (2014) Preparation and characterization of bionanocomposite films based on potato starch/halloysite nanoclay. Int J Biol Macromol 67:458–462
61. Nguyen TP (2013) Applications of polymer-based nanocomposites. In: Thomas S, Malhotra SK, Goda K, Sreekala MS (eds) Polymer composites. Wiley, Germany, pp 249–277
62. Shah AA, Hasan F, Hameed A, Ahmed S (2008) Biological degradation of plastics: a comprehensive review. Biotechnol Adv 26(3):246–265
63. Lucas N, Bienaime C, Belloy C, Queneudec M, Silvestre F, Nava-Saucedo JE (2008) Polymer biodegradation: mechanisms and estimation techniques. Chemosphere 73(4):429–442
64. Thakore IM, Desai S, Sarawade BD, Devi S (2001) Studies on bioderadability, morphology and theromechanical properties of LDPE/modified starch blend. Eur Polym J 37:151–160
65. Silvestre C, Duraccio D, Cimmino S (2011) Food packaging based on polymer nanomaterials. Prog Polym Sci 36(12):1766–1782

66. Arora A, Padua GW (2010) Review: nanocomposites in food packaging. J Food Sci 75(1): R43–R49
67. Seymour GB, Tucker GA (1993) Avocado. In: Seymour GB, Taylor JE, Tucker GA (eds) Biochetnistry of fruit ripening. Springer, Malaysia, pp 53–82
68. Brady CJ (1993) Stone fruit. In: Seymour GB, Taylor JE, Tucker GA (eds) Biochetnistry of fruit ripening. Springer, Malaysia, pp 379–397
69. Hu Q, Fang Y, Yang Y, Ma N, Zhao L (2011) Effect of nanocomposite-based packaging on postharvest quality of ethylene-treated kiwifruit (*Actinidia deliciosa*) during cold storage. Food Res Inter 44(6):1589–1596

Chapter 3
Morphological Structures, Mechanical, Thermal and Optical Properties of PVA/HNT Bionanocomposite Films

Abstract The effect of material composition and nanofiller contents on mechanical, thermal and optical properties along with morphological structures was evaluated in this chapter. The presence of glycerol (GL) as a typical plasticiser reduced Young's modulus and tensile strength of polyvinyl alcohol (PVA)/GL blends, as well as improved the elongation at break when compared with those of neat PVA films. This trend was completely opposite to the effect of starch (ST) for PVA/ST blends. Moreover, tensile strength, Young's modulus and thermal properties in term of melting temperature (T_m), decomposition temperatures and weight loss of PVA/ST/GL/halloysite nanotube (HNT) bionanocomposites were enhanced significantly with the incorporation of 0.25–1 wt% HNTs due to good nanofiller dispersion to form intercalated bionanocomposite structures. Such properties of bionanocomposite films appeared to decline beyond 1 wt% HNTs due to their typical agglomeration, as evidenced by scanning electron microscopy (SEM) and atomic force microscopy (AFM) in spite of being still better than those of PVA counterparts.

Keywords Mechanical properties · Thermal properties · Intercalated structures · Transparency · Filler agglomeration

Although biodegradable polymers are widely available, relatively cheap, non-toxic, highly reactive, biocompatible and biodegradable along with acceptable strength, they still have very narrow applications due to their weak mechanical properties and thermal stability, as well as poor barrier properties [1–3]. Food packaging requires a combination of science, technology and art to provide physical protection of products in order to maintain their quality and shelf life by minimising the permeability at the least price [2, 4, 5]. The most popular scenario to overcome these limitations is to develop new polymer nanocomposite systems for packaging applications [6, 7]. Polymer nanocomposites offer the potential improvements in material properties like mechanical, thermal, optical and barrier properties at relatively low nanofiller contents [8]. A number of molecular changes can happen in nanocomposites due to the interactions between nanofillers with high surface areas

© Springer Nature Singapore Pte Ltd. 2020
Z. W. Abdullah and Y. Dong, *Polyvinyl Alcohol/Halloysite Nanotube Bionanocomposites as Biodegradable Packaging Materials*,
https://doi.org/10.1007/978-981-15-7356-9_3

and polymer matrices leading to the "non-classical" response of these nanocomposites reflected on the change of bulk material properties [9]. Consequently, an explicit investigation on morphological structures of nanocomposites is essential to establish their good structure–property relationship.

3.1 Morphological Structures

3.1.1 SEM Observation

Morphological structures of as-received HNT powders, neat PVA and PVA blends are illustrated in Fig. 3.1. Whereas, morphological structures of PVA/ST/GL/HNT bionanocomposite films at different HNT contents are depicted in Fig. 3.2. A tubular structure of HNTs has been clearly observed in Fig. 3.1a. Moreover, typical HNT agglomeration is also noticeable as HNT nanoparticles naturally tend to agglomerate due to their weak van der Waals interactions. As seen from Fig. 3.1b, fracture surfaces of neat PVA appear to be smooth, brittle with the limited elongation and the coexistence of crystalline and amorphous phases in the films, as mentioned earlier. However, more ductile material behaviour is clearly identified for PVA/GL blends owing to the GL plasticisation effect shown in Fig. 3.1c [10]. Whereas, fracture surfaces of PVA/ST blend films tend to be much rougher and more brittle with the addition of ST (Fig. 3.1d) [11, 12]. The further inclusion of GL depicted in Fig. 3.1e is shown to diminish such brittleness nature. The constituents of PVA/ST/GL blends demonstrate good compatibility with one another without apparent phase separation, which is in good agreement with the results obtained by Wu et al. [13]. Their SEM results reveal that PVA/ST blends have homogenous and dense film structures in the presence of GL and citric acid with better binding improvements between different material components [13].

This identical compatibility occurs again in bionanocomposite films. In other words, the presence of HNTs does not affect the compatibility of PVA/ST/GL blends. Low HNT contents at 0.25 and 1 wt% yield homogeneous HNT dispersion in bionanocomposites (as circled in white colour) with typical wavy-line structures (as circled in red colour), as demonstrated in Fig. 3.2a–c. Such a phenomenon has also been identified in the previous study carried out by Khoo et al. [15] for PVA/chitosan/HNT nanocomposites at the HNT contents of 0.25 and 0.5 wt%, respectively. These wavy-line structures are believed to result in the enhancement of tensile strength of such bionanocomposite films when compared with that of their polymer blends alone due to an increase in tearing strength. On the other hand, a clear sign of HNT agglomeration has been detected at the HNT contents of 3 and 5 wt% in PVA/ST/GL/HNT bionanocomposites, as illustrated in Fig. 3.1d, e accordingly, which is consistent with previous work [15]. As such, the tendency of HNT agglomeration increases with increasing their weight fraction, as evidenced by Gaaz et al. [16].

Energy-dispersive spectroscopic (EDS) tests were performed for bionanocomposite films at the HNT contents of 0.25 and 3 wt% to highlight two typical

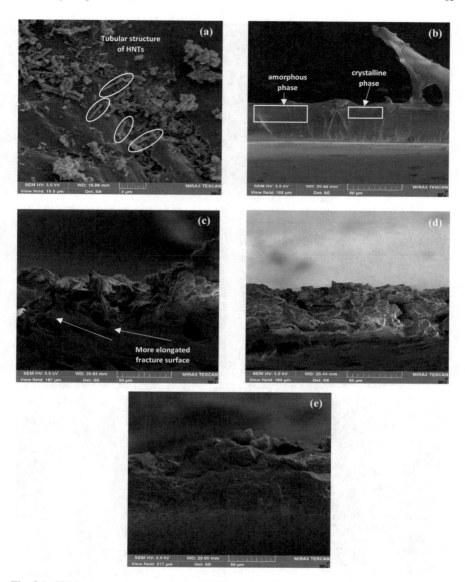

Fig. 3.1 SEM micrographs: **a** as-received HNT powders, **b** pure PVA, **c** PVA/GL blends, **d** PVA/ST blends and **e** PVA/ST/GL blends. Images taken from [14] with the copyright permission from Springer

conditions of well-dispersed HNTs and HNT agglomerates, respectively, Fig. 3.3. The presence of Al and Si elements in the chemical composition of bionanocomposite films is associated with dispersed HNTs within blend matrices as Al and Si are the major elements of HNT chemical structure (i.e. $Al_2Si_2O_5(OH)_4 \cdot nH_2O$) in

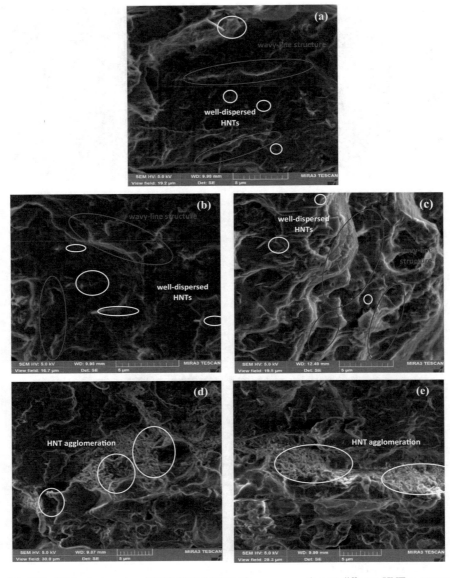

Fig. 3.2 SEM micrographs of PVA/ST/GL/HNT bionanocomposites at different HNT contents: **a** 0.25 wt%, **b** 0.5 wt%, **c** 1 wt%, **d** 3 wt% and **e** 5 wt%. Images taken from [14] with the copyright permission from Springer

good accordance with Dong et al. [17]. The presence of carbon and oxygen elements is attributed to the existence of PVA/ST/GL blends with additional carbon contents derived from carbon-coating sample layers.

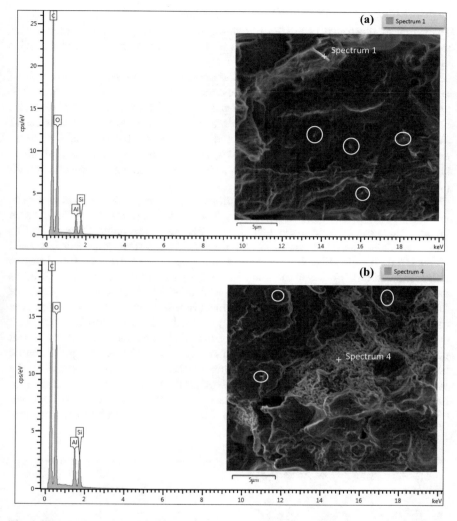

Fig. 3.3 EDS spectra of PVA/ST/GL/HNT bionanocomposites at two typical HNT contents: **a** 0.25 wt% and **b** 3 wt%. Circled areas indicate dispersed HNT particles within polymer blend matrices. Images taken from [14] with the copyright permission from Springer

3.1.2   AFM Characterisation

The tubular morphology of as-received HNTs was observed again from the AFM images shown in Fig. 3.4a. Moreover, the average dimensions of as-received HNTs were calculated as well based on the measurements of 173 individual particles despite the difficulty to overcome HNT agglomeration. As-received HNTs have an average diameter $D = 18.63 \pm 0.52$ nm and an average length $L = 730.75 \pm 13.40$ nm, as illustrated in Fig. 3.4b. Consequently, corresponding average aspect ratio (L/D) of as-received HNTs becomes 39.22, which is in good accordance with other studies [18, 19].

AFM images of PVA/ST/GL blends and corresponding bionanocomposite films at different HNT contents clearly demonstrate three major areas including black areas to represent amorphous phase of polymer blend matrices, light and dark brown areas for their crystalline phase and interphase between polymer blend matrices and HNTs, as well as white and yellow areas for fully and partially embedded HNTs within polymer matrices, respectively (see Fig. 3.5), which is based on diverse mechanical properties of each surface area in good agreement with previous literatures [21, 22].

The variation of surface mechanical properties can produce different contrasts under AFM corresponding to each area. The AFM technique in a Peak Force tapping mode can be used to evaluate these variations in mechanical properties like elastic modulus, adhesion and dissipation of a given surface area. For example, PVA/ST/GL blends (Fig. 3.6a) were scanned along A-A1 section (Fig. 3.6b) to reflect the variation of elastic modulus for each area along the scanning section. As exhibited in Fig. 3.5c, the light areas have a relatively high elastic modulus of 103.7 ± 9.3 MPa related to that of crystalline phases. Whereas, the dark areas give rise to a much lower elastic modulus of 21.2 ± 4.1 MPa for amorphous phases.

Furthermore, bionanocomposite films at the HNT contents up to 1 wt% reflect good HNT dispersion within polymer matrices, beyond which HNT agglomeration becomes more pronounced. These observations have also been supported by the evaluation of surface roughness of PVA/ST/GL blend films and their bio-nanocomposite films. The surface roughness of bionanocomposite films increases with increasing the HNT content from 0 to 5 wt% in a more and less linear manner

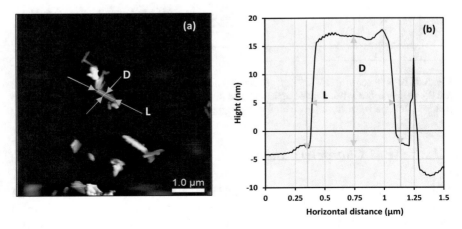

Fig. 3.4 **a** AFM image of as-received HNTs and **b** typical height section profile of individual HNT. Images taken from [20] with the copyright permission from Elsevier

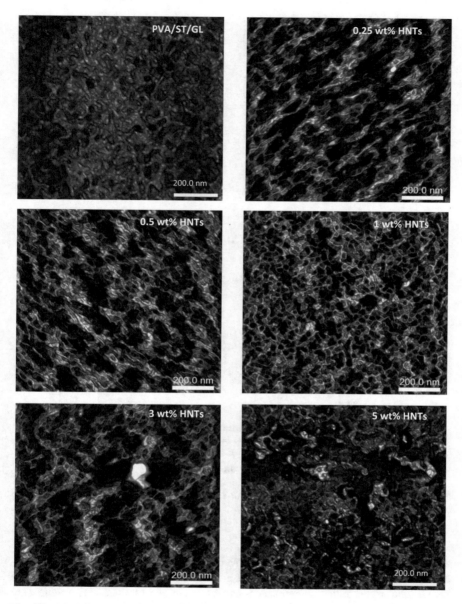

Fig. 3.5 AFM images of PVA/ST/GL blends and corresponding bionanocomposite films at different HNT contents. Images taken from [20] with the copyright permission from Elsevier

from 13.83 to 75.80 nm. This trend appears to be less pronounced beyond 1 wt% HNTs, as shown in Fig. 3.7. The lower surface roughness of bionanocomposite films with the HNT inclusion up to 1 wt% can be attributed to uniform HNT

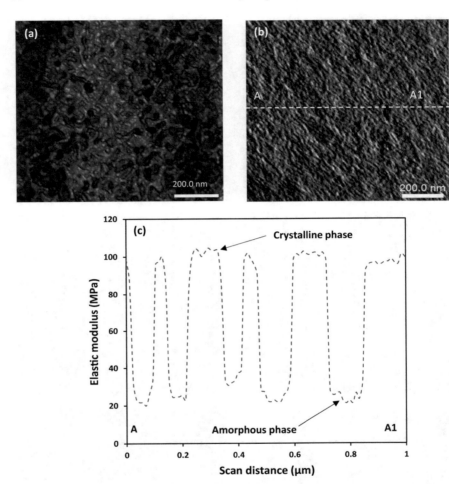

Fig. 3.6 a 2D AFM mapping image of PVA/ST/GL blend films, **b** elastic modulus mapping image and **c** elastic modulus versus scan distance curve for typical section A-A1 in (**b**). Images taken from [20] with the copyright permission from Elsevier

dispersion within polymer matrices, as well as homogenous structures built up by effective intermolecular bonding between different components [23]. It is well known that aspect ratios of nanofillers can be remarkably reduced with increasing the nanofiller content [24–26], particularly resulting from their agglomeration at high content levels. This phenomenon has been proven experimentally in this study by determining the aspect ratios of HNTs within bionanocomposite films at different HNT contents. HNT aspect ratios decreased in a monotonic manner from 39.27 to 14.87 with increasing HNT contents from 0.25 to 5 wt% illustrated in Fig. 3.7. In short, the surface roughness of bionanocomposite films and embedded

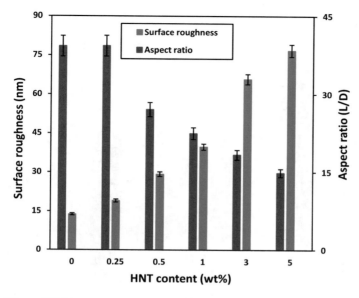

Fig. 3.7 Effect of HNT content on surface roughness and aspect ratio of embedded HNTs in PVA/ST/GL/HNT bionanocomposite films. The figure is redrawn from Ref. [20] with the copyright permission from Elsevier

HNT aspect ratios completely reflect an opposite trend. In other words, increasing the HNT content could further increase the surface roughness of bionanocomposite films along with the reduction in HNT aspect ratios within their films accordingly.

3.2   Mechanical Properties

Mechanical properties of neat PVA, PVA/GL, PVA/ST and PVA/ST/GL blends are summarised in Fig. 3.8. Plasticised PVA films with 30 wt% GL were prepared to improve the elasticity and reduce the brittleness of neat PVA. Significant reductions in tensile strength and Young's modulus of PVA/GL blend films have been found to be 77.95 and 96.5%, respectively, as compared with those of neat PVA. Nonetheless, the elongation at break is shown to be increased drastically by 321.09% accordingly owing to typical GL plasticisation effect to improve the movement of polymeric chains, and thus increase free volume resulting in ductile fracture surfaces in the SEM results obtained. Moreover, PVA/ST blends have much lower tensile strength and elongation at break than those of neat PVA by 28.32 and 51.66%, respectively, despite an increase in Young's modulus by 52.68% as opposed to that of neat PVA. This trend is ascribed to amorphous nature and inherent brittleness of ST leading to the improvement of stiffness at the expense

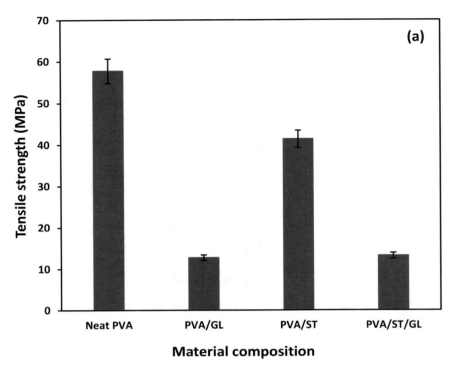

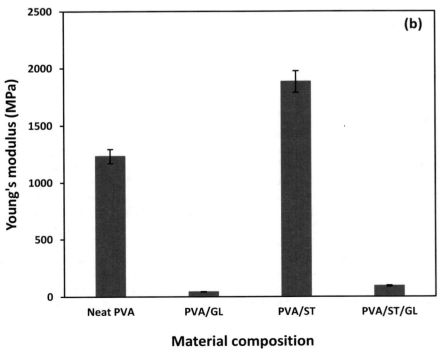

Fig. 3.8 Mechanical properties of neat PVA, PVA/GL, PVA/ST and PVA/ST/GL blends: **a** tensile strength, **b** Young's modulus and **c** elongation at break

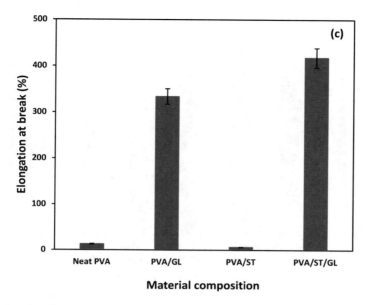

Fig. 3.8 (continued)

of flexibility in good agreement with Azahari et al. [27] and Ramaraj [28], as well as the rough fractured surfaces depicted in SEM images of this study. Additionally, PVA/ST/GL blends achieve the improvements in tensile strength and Young's modulus by 4.0 and 119.67% relative to those of PVA/ST blends though still lower than those of neat PVA and PVA/ST blends. Compared with both neat PVA and PVA/ST blends, PVA/ST/GL blends possess excellent elongation at break, which is regarded as one of key mechanical properties in relation to nanocomposite manufacturing processes. According to Wu et al. [13], the elongation at break may vary as a function of the GL content. GL with hydroxyl groups can form hydrogen bonds with polymers via the interaction with hydroxyl and carboxyl groups as well to effectively improve the free volume of the material system by reducing intermolecular forces and change polymer blends from typical brittle nature to good ductility [13]. Overall, these findings are quite consistent with the morphological structures displayed in SEM analysis.

PVA/ST/GL/HNT bionanocomposite films at the HNT contents of 0.25 and 0.5 wt% have higher tensile strength than that of PVA/ST/GL blends with an increasing level by 20.0 and 3.4%, respectively (see Fig. 3.9). Moreover, the tensile strength of bionanocomposite films declines significantly beyond 1 wt% HNTs. This trend can be interpreted by typical HNT agglomeration beyond 1 wt% to induce the detrimental effect on interfacial bonding between polymer blend matrices and nanofillers for less effective load transfer, which has been clearly demonstrated in the SEM and AFM results. Similar trends were also reported by Sadhu et al. [29] for PVA/ST/Cloisite 30B clay nanocomposite films and Tang et al. [30] for PVA/ST/nanoSiO$_2$ nanocomposite films. Moreover, Young's moduli of

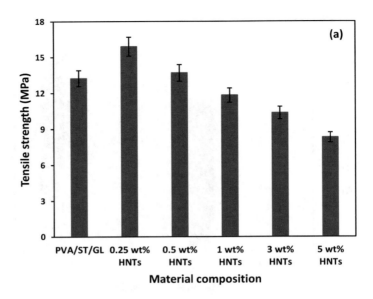

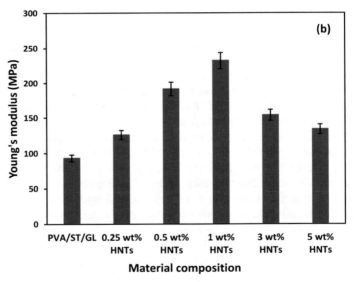

Fig. 3.9 Mechanical properties of PVA/ST/GL blends and corresponding bionanocomposite films at different HNT contents: **a** tensile strength, **b** Young's modulus and **c** elongation at break. Figures taken from [14] with the copyright permission from Springer

bionanocomposite films are enhanced linearly with increasing the HNT content from 0 to 1 wt% up to 148.97% due to the inherent toughness of HNTs [31]. Although Young's moduli of bionanocomposite films reinforced with 3 and 5 wt%

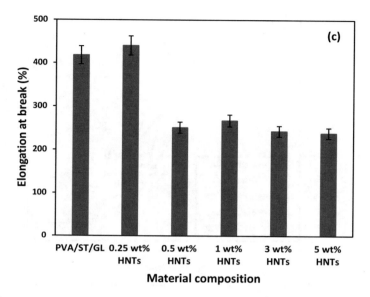

Fig. 3.9   (continued)

HNTs decline as compared to those with lower HNT contents, it is still higher than that of PVA/ST/GL blends. Accordingly, the elongation at break decreases remarkably with increasing the HNT content beyond 0.25 wt% due to the mobility restriction of polymeric chains in the presence of nanofillers [30]. Filler agglomeration arising from poor dispersion within polymer matrices can be the main reason associated with the similar behaviour, as evidenced by Heidarian et al. [32] for PVA/ST/cellulose nanofibril (CNF) nanocomposite films and Cano et al. [33] for PVA/ST/cellulose nanocrystal (CNC) nanocomposite films. Overall, the associated improvements in tensile strength and Young's modulus are related to nanofiller dispersion and nanofiller content, while the elongation at break is mainly influenced by the plasticiser content in polymer blends, which decreases with increasing the nanofiller content as well.

3.3   Thermal Properties

Thermogravimetric analysis (TGA) curves and derivative thermogravimetric (DTG) curves of neat PVA and ST, as well as their blends are shown in Fig. 3.10a, b, respectively. Additionally, the data of these curves are summarised in Table 3.1 accordingly. The thermal stability of PVA/GL blends is relatively low in comparison with that of neat PVA since GL plasticisation effect can weaken intermolecular and intramolecular forces with the improvement of chain mobility leading to an increase in both heat and mass transfer. Similarly, Mohsin et al. [34]

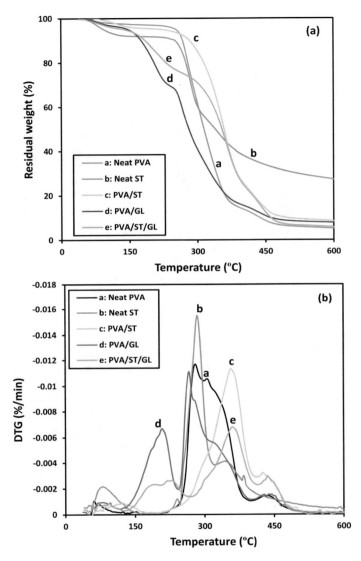

Fig. 3.10 a TGA curves and **b** DTG curves of neat PVA, ST and their polymer blends. Figures taken from [14] with the copyright permission from Springer

concluded that the plasticisers had low molecular weight compared with those of polymers, and easily penetrated between polymeric molecules to reduce their interactions. As such, chain mobility and free volume are increased resulting in the final reduction in their thermal stability. Such findings are clearly demonstrated with decreasing the decomposition temperatures of PVA/GL blends by 104.7, 32.8

Table 3.1 TGA and DTG data for neat materials, PVA blends and PVA/ST/GL/HNT bionanocomposites at different HNT contents

Sample	$T_{5\%}$ (°C)	$T_{50\%}$ (°C)	$T_{90\%}$ (°C)	T_{d1} (°C)	T_{d2} (°C)	T_{d3} (°C)	T_{d4} (°C)
Neat PVA	253.2	314.3	436.9	76.1	–	282.1	432.0
Neat ST	94.1	338.6	–	81.1	–	285.3	349.8
As-received HNTs	–	–	448.9	–	–	–	497.0
PVA/GL	148.5	281.1	429.1	80.9	211.0	267.7	440.0
PVA/ST	221.2	358.5	494.7	126.9	–	359.2	428.4
PVA/ST/GL	135.3	347.1	460.4	90.3	211.3	350.3	438.2
PVA/ST/GL/0.25 wt % HNTs	139.5	359.7	460.8	91.7	210.1	364.4	439.6
PVA/ST/GL/0.5 wt% HNTs	144.1	356.1	466.8	92.9	200.1	364.6	443.3
PVA/ST/GL/1 wt% HNTs	155.8	355.8	468.9	96.7	230.0	365.0	446.1
PVA/ST/GL/3 wt% HNTs	153.3	345.9	468.0	91.4	195.5	362.4	443.0
PVA/ST/GL/5 wt% HNTs	137.5	347.7	–	93.2	190.5	359.2	445.8

Table taken from [14] with the copyright permission from Springer

and 7.8 °C at the weight losses of 5, 50 and 90%, respectively, as opposed to that of neat PVA, which are represented by corresponding $T_{5\%}$, $T_{50\%}$ and $T_{90\%}$. In particular, higher thermal stability of PVA/ST blends compared with that of neat PVA can be explained by inherent ST structures of thermally resistive cyclic hemiacetal [35]. As a result, the $T_{50\%}$ and $T_{90\%}$ of PVA/ST blends appear to be increased by 44.2 and 57.8 °C, respectively, as opposed to those of neat PVA. Furthermore, the thermal stability of PVA/ST/GL blends is not comparable to those of PVA and ST alone in that such blends have complex polymeric components with plenty of hydroxyl and carboxyl groups that can be actively interacted for generating hydrogen bonds. In short, thermal properties of PVA/ST/GL appear to be between those of PVA/GL blends and PVA/ST blends.

On the other hand, the thermal stability of PVA/ST/GL/HNT bionanocomposites is even better than that of PVA/ST/GL blends, as evidenced by increasing decomposition temperatures and decreasing the weight loss illustrated in Fig. 3.11a, b. At the HNT content of 1 wt%, $T_{5\%}$, $T_{50\%}$ and $T_{90\%}$ of bionanocomposites are mostly the highest among other bionanocomposites, which are increased by 20.5, 8.7 and 8.5 °C, respectively, when compared with those of PVA/ST/GL blends, Table 3.1. Notwithstanding that the thermal stability of bionanocomposites with the inclusion of 3 and 5 wt% HNTs slightly declines due to the HNT agglomeration, it still appears to be better than those of PVA/ST/GL blends. The inherent characteristic of HNTs as a barrier material against heat and mass transfer is clearly shown because hollow tubular structures enable to trap volatile molecules to delay the

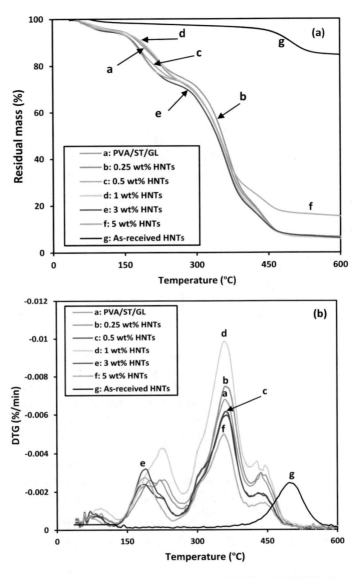

Fig. 3.11 **a** TGA curves and **b** DTG curves for as-received HNTs, PVA/ST/GL blends and their corresponding bionanocomposites at different HNT contents. Figures taken from [14] with the copyright permission from Springer

mass transfer during the thermal decomposition process, which is deemed as the major reason for improving the thermal stability of bionanocomposite films [36].

DTG curves of neat PVA and ST shown in Fig. 3.10b indicate clear decomposition temperatures (T_{d3}) with associated sharp peaks taking place at 282.1 and

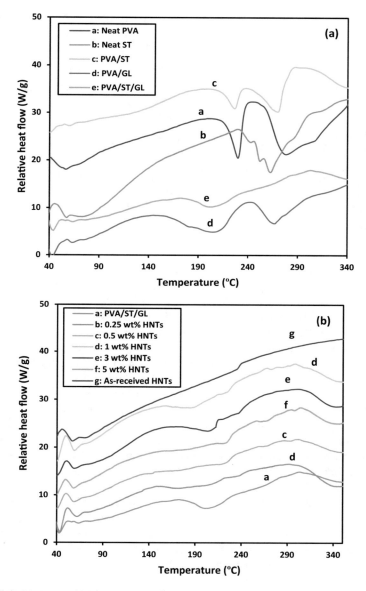

Fig. 3.12 DSC thermograms: **a** neat PVA and ST and corresponding blends and **b** as-received HNTs, PVA/ST/GL blend and their corresponding bionanocomposites at different HNT contents. Figures taken from [14] with the copyright permission from Springer

285.3 °C, respectively, which is related to the decomposition and dehydration of hydrogen bonds. Additionally, water loss and carbonisation of organic molecules give rise to the other two unclear decomposition temperatures detected below 100 °

Table 3.2 DSC data for neat materials, PVA blends and PVA/ST/GL/HNT bionanocomposites at different HNT contents

Sample	T_g (°C)	T_c (°C)	T_m (°C)	ΔH_c (J/g)	ΔH_m (J/g)	X_c (%)
Neat PVA	70.71	229.90	244.69	8.64	30.14	15.14
Neat ST	80.49	–	263.30	–	30.52	–
PVA/GL	47.75	237.02	229.01	19.37	27.08	5.42
PVA/ST	74.12	269.7	288.41	10.18	21.29	7.82
PVA/ST/GL	56.88	204.28	290.04	6.45	20.32	9.76
PVA/ST/GL/0.25%HNTs	66.83	203.33	303.11	6.38	28.8	16.19
PVA/ST/GL/0.5%HNTs	66.56	200.39	301.72	5.25	25.46	14.98
PVA/ST/GL/1%HNTs	65.88	192.17	300.34	8.19	26.38	14.23
PVA/ST/GL/3%HNTs	65.45	202.80	304.36	7.41	23.46	11.65
PVA/ST/GL/5%HNTs	64.22	200.39	304.52	3.72	18.91	11.15

Table taken from [14] with the copyright permission from Springer

C (T_{d1}) and above 400 °C (T_{d4}), respectively, as confirmed in previous work [37, 38]. A new decomposition temperature (T_{d2}) at approximately 200 °C is attributed to the evaporation of volatile materials for PVA/GL blends. In a similar manner, PVA/ST blends have comparable decomposition temperatures to that of neat PVA with a slight increase in T_{d1} associated with the moisture content within ST structures. PVA/ST/GL blends possess four decomposition temperatures similar to that of PVA/GL blends, namely T_{d1} below 100 °C for water loss, T_{d2} at about 225 ° C due to the presence of GL as a plasticiser, T_{d3} at approximately 360 °C arising from the decomposition/dehydration of hydrogen bonds, as well as T_{d4} at about 435 °C in relation to the carbonisation of organic molecules [39, 40]. All decomposition temperatures of bionanocomposites are enhanced when compared with those of corresponding polymer blends, which is ascribed to the incorporation of HNTs with a single decomposition temperature detected at 490.0 °C due to their higher thermal stability (Fig. 3.11b). T_{d1}, T_{d2}, T_{d3} and T_{d4} of bionanocomposites reinforced with 1 wt% HNTs are increased by 6.4, 18.7, 14.7 and 7.8 °C, respectively, when compared with those of PVA/ST/GL blends. However, they appear to decline slightly at HNT contents of 3 and 5 wt% owing to HNT agglomeration at high content levels. In comparison, at the low HNT contents between 0.25 and 1 wt%, the enhancement of thermal stability of bionanocomposites is believed to be associated with good HNT dispersion resulting in increasing the heterogeneity of polymer blends, which is consistent with Priya et al. [41] for PVA/ST/CNF nanocomposites and Sadhu et al. [42] for PVA/ST/Cloisite 30B clay nanocomposites.

Differential scanning calorimetry (DSC) curves of neat PVA, neat ST and their blends are demonstrated in Fig. 3.12a. Moreover, DSC curves of as-received HNTs, PVA/ST/GL blends and their bionanocomposites are exhibited in Fig. 3.12b along with associated thermal properties listed in Table 3.2. Glass transition temperature (T_g) and T_m of neat PVA have been determined to be 70.7 and 244.6 °C,

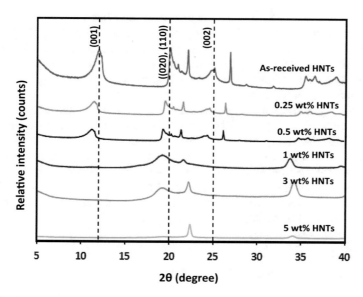

Fig. 3.13 XRD patterns of as-received HNTs and PVA/ST/GL/HNT bionanocomposites at different HNT contents. Figure taken from [14] with the copyright permission from Springer

Table 3.3 d-spacing values of as-received HNTs and embedded HNTs in PVA/ST/GL bionanocomposites

XRD sample	2θ	d_{001} (nm)	2θ	$d_{020/110}$ (nm)	2θ	d_{002} (nm)
As-received HNTs	12.12	0.73	20.01	0.44	24.98	0.35
PVA/ST/GL/0.25 wt% HNTs	11.54	0.76	19.22	0.46	24.00	0.37
PVA/ST/GL/0.5 wt% HNTs	11.21	0.79	19.02	0.46	23.57	0.38
PVA/ST/GL/1 wt% HNTs	–	–	18.95	0.47	–	–
PVA/ST/GL/3 wt% HNTs	–	–	18.40	0.48	–	–
PVA/ST/GL/5 wt% HNTs	–	–	–	–	–	–

Table taken from [14] with the copyright permission from Springer

respectively, which is in good agreement with previous studies [43–45]. Whereas, the T_g and T_m of neat ST are relatively high reaching 80.4 and 263.3 °C, respectively. The good compatibility between PVA, ST and GL is indicated by a single T_g for all PVA blends, which is supported by the SEM observation mentioned earlier. T_g and T_m of PVA/ST blends are improved by 3.4 and 43.7 °C, respectively, as opposed to those of neat PVA, which is ascribed to better stiffening effect of hydrogen bonds between PVA and ST with the addition of ST to induce new hydroxyl groups [28, 46]. Conversely, T_g and T_m of PVA/GL blends are decreased by 23.0 and 15.6 °C, respectively, relative to those of PVA due to aforementioned GL plasticisation effect.

On the other hand, T_g and T_m of bionanocomposites are increased with the inclusion of HNTs at different rates as opposed to those of PVA/ST/GL blends. For instance, the higher increasing rates of T_g by 9.95 °C at 0.25 wt% HNTs and T_m by 14.48 °C at 5 wt% HNTs could be associated with high thermal stability of HNTs. It is well known that HNTs can act as heterogeneous nucleating agents when incorporated into polymer matrices in bionanocomposite material systems [36]. Hence, the crystallisation temperature (T_c) of bionanocomposites tends to be decreased with the addition of HNTs, while the related crystallisation rate (X_c) is enhanced with the incorporation of HNTs particularly at HNT contents between 0.25 and 1 wt% when compared with those of PVA/ST/GL blends counterparts. Furthermore, the melting enthalpy (ΔH_m) of bionanocomposites is improved as well with the addition of HNTs because of thermal stability and barrier action of HNTs against heat and mass transfer [36, 47], which also coincides with Tee et al. [48] in PVA/ST/montmorillonite (MMT) nanocomposites.

3.4 X-Ray Diffraction (XRD) Analysis

The XRD patterns of as-received HNTs and PVA/ST/GL/HNT bionanocomposites at different HNT contents are presented in Fig. 3.13, and corresponding d-spacing values are summarised in Table 3.3. As-received HNTs have three major peaks at diffraction angles 2θ of 12°, 20° and 25°, respectively, corresponding to (001), ((020), (110)) and (002) crystal planes, respectively. Based on Bragg's law, their resulting d-spacing values have been calculated to be approximately 0.73, 0.44 and 0.35 nm, respectively. PVA/ST/GL/HNT bionanocomposites reflect a slight peak shift from 12.12° to 11.54° and 11.21° with the addition of 0.25 and 0.5 wt% HNTs, respectively. These slight changes are considered as minor intercalated clay structures with respect to (001) peak, which is in good agreement with Dong et al. [49] in polylactic acid (PLA)/HNT nanocomposites. Furthermore, the other peaks at ((020), (110)) and (002) have a similar trend for intercalated structures with a slight peak shift as well. Nonetheless, XRD peaks completely disappear for bionanocomposites beyond 1 wt% HNTs, which could be principally associated with the combination of well-dispersed and highly agglomerated HNTs in the disordered orientation.

3.5 Fourier Transformation Infrared (FTIR) Analysis

Neat polymers consisting of PVA, ST and GL, as-received HNTs, PVA blends and PVA/ST/GL/HNT bionanocomposites at different HNT contents were examined via FTIR analysis. As shown in Fig. 3.14, many peaks have been observed from these spectra, and each peak is related to predefined functional groups in polymeric chains. For neat PVA, ST and GL, a distinct peak detected at 3200–3300 cm^{-1} is associated with the O–H stretching due to the strong molecular hydrogen bonding.

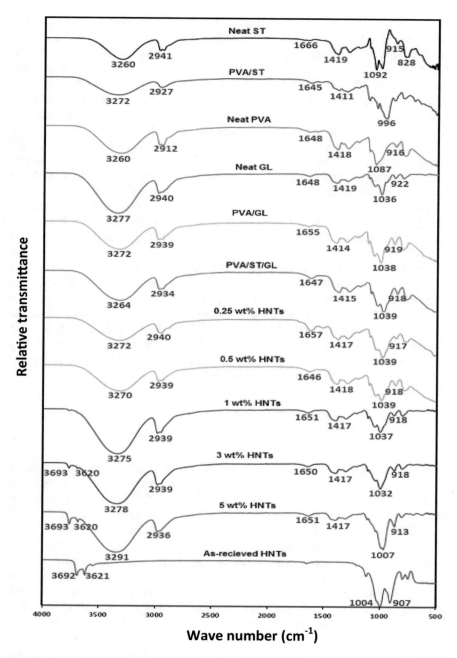

Fig. 3.14 FTIR spectra of neat PVA, ST, GL, as-received HNTs, PVA blends and PVA/ST/GL/ HNT bionanocomposites at different HNT contents. Figure taken from [14] with the copyright permission from Springer

The other apparent peaks identified in range of 2900–2950, 1600–1650, 1414–1420 and 1000–1090 cm^{-1} are due to C–H, bonding water, CH$_2$ groups, C–C–C and C–O stretchings, respectively, in good accordance with previous findings [13, 33, 39, 50]. PVA/ST, PVA/GL and PVA/ST/GL blends have similar peaks with some changes taking place in their intensity and wavelength of O–H stretching due to their good compatibility to increase hydrogen bonding [12]. Nonetheless, slight changes in other peaks are manifested, possibly resulting from intermolecular and intramolecular bonding between blend components [50].

As-received HNTs show two distinct peaks at 3692 and 3621 cm^{-1}, which are assigned to the O–H stretching vibration. According to Gaaz et al. [16], the first peak is related to inner surface O–H groups in connection with the aluminium centred sheet, while the second peak is ascribed to inner O–H groups. As-received HNTs have other peaks at 1004 and 907 cm^{-1}, which are assigned to Si–O and Al–OH stretchings, respectively [17]. The O–H stretching of as-received HNTs disappears when embedded within polymer blend matrices in bionanocomposite films at HNT contents of 0.25–1 wt%. This phenomenon can be interpreted by the emerging O–H stretching of HNTs within PAV/ST/GL blend matrices, arising from good HNT dispersion at low nanofiller contents in bionanocomposite films. On the other hand, the O–H stretching is identified again for bionanocomposites with the inclusion of 3 and 5 wt% HNTs with a clear sign of HNT agglomeration, as confirmed in the XRD results. Moreover, with increasing the HNT content, Si–O and Al–OH stretchings become more pronounced, while other peaks identified at 1650 and 1417 cm^{-1} suggest typical HNT-matrix interaction at the specific sites for bonding water and CH$_2$ stretching, respectively.

3.6 UV-Vis Spectra

Film transparency in term of light transmittance ($T\%$) is an important material feature in food packaging applications. Consequently, neat PVA, PVA blends and PVA/ST/GL/HNT bionanocomposite films at different HNT contents were evaluated under UV-visible light spectra, as well as Curtin University logo was observed through all these films, as displayed in Fig. 3.15a, b, respectively.

The high degree of crystallinity for neat PVA leads to a high $T\%$ range of 99–100% in good agreement with previous results [51, 52], as summarised in Table 3.4. PVA/GL, PVA/ST and PVA/ST/GL blend films have relatively high $T\%$ in range of 95.23–98.72% because of the good component miscibility consistent with the SEM results. Although these blends have good $T\%$, it is still lower than that of neat PVA because the degree of crystallinity for these films becomes relatively low as opposed to that of neat PVA, which has been clearly demonstrated in our DSC results (see Table 3.2). As shown from Curtin University's logo, there is no great difference in clarity between neat PVA and their blends.

The incorporation of HNTs within PVA/ST/GL blend matrices reduces $T\%$ of bionanocomposite films when further compared with that of corresponding blends

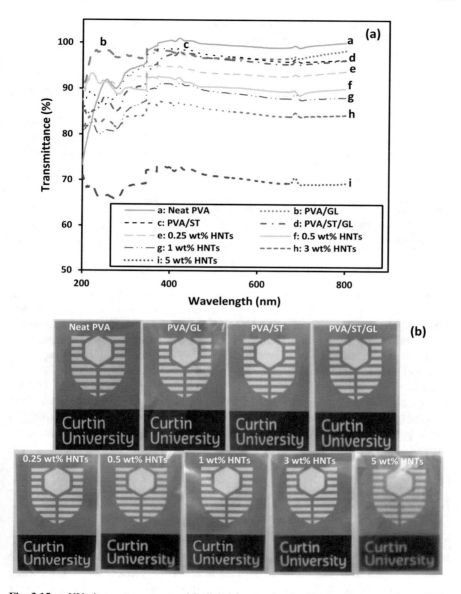

Fig. 3.15 a UV-vis spectra curves and **b** digital images for the film transparency of neat PVA, PVA blends and PVA/ST/GL/HNT bionanocomposite films at different HNT contents [53]

alone. Regardless of the wavelength used, the reduction of $T\%$ for bionanocomposite films increases linearly with increasing the HNT contents from 0.25 to 5 wt %, which is attributed to an increase in surface roughness of bionanocomposite films particularly at HNT contents of 3 and 5 wt%. As such, the number of light

Table 3.4 Visible light transmittance data for neat PVA, PVA blends and PVA/ST/GL/HNT bionanocomposite films at different HNT contents

Material composition	$T\%$ at a visible wavelength range			
	400 nm	500 nm	600 nm	700 nm
Neat PVA	100.00 ± 0.091	99.67 ± 0.30	99.13 ± 0.72	98.99 ± 0.48
PVA/GL	98.15 ± 0.56	96.66 ± 0.71	96.50 ± 0.24	96.67 ± 0.66
PVA/ST	98.72 ± 0.51	97.29 ± 0.18	96.43 ± 0.84	95.76 ± 0.69
PVA/ST/GL	97.43 ± 0.75	96.59 ± 0.86	95.73 ± 0.55	95.23 ± 0.72
PVA/ST/GL/0.25 wt% HNTs	89.79 ± 0.73	93.38 ± 0.67	92.91 ± 0.48	92.81 ± 0.50
PVA/ST/GL/0.5 wt% HNTs	85.32 ± 0.52	90.36 ± 0.77	89.79 ± 0.21	88.71 ± 0.17
PVA/ST/GL/1 wt% HNTs	81.02 ± 0.93	89.35 ± 0.97	88.38 ± 0.74	87.52 ± 0.89
PVA/ST/GL/3 wt% HNTs	86.90 ± 0.30	85.43 ± 0.79	84.48 ± 0.73	83.87 ± 0.27
PVA/ST/GL/5 wt% HNTs	72.76 ± 0.58	70.90 ± 0.80	69.87 ± 0.20	68.90 ± 0.31

scattering sites can be increased accordingly [54], as reflected clearly by reducing the visibility of Curtin University's logo particularly with the inclusion of 5 wt% HNTs. These results are consistent with the AFM data in term of surface roughness. Moreover, such a decreasing trend of $T\%$ for PVA/ST/GL/HNT bionanocomposite films has been confirmed by other studies in PVA/ST/nanoSiO$_2$ nanocomposites [30] and PVA/ST/ZnO nanocomposites [50]. On the other hand, Lee et al. [55] considered the reduction of $T\%$ was advantageous when analysing the UV-vis spectra for chitosan/clove essential oil/HNT nanocomposite films used for food packaging applications because UV-barrier properties were improved leading to better protection of foodstuffs against nutrient loss, decolourisation and lipid oxidation.

3.7 Summary

This chapter covers the effects of HNT content on morphological structures, mechanical, thermal and optical properties of PVA/ST/GL/HNT bionanocomposite films in comparison with those of neat PVA and PVA blends. Overall, blending PVA with GL substantially increases the elongation at break as compared with neat PVA at the expense of tensile strength and Young's modulus due to the improvement of polymeric chain mobility and free volume. A completely opposite behaviour has been observed when blending PVA with ST due to the latter's inherent brittleness. Moreover, mechanical properties of PVA/ST/GL blends have been found to range between those of PVA/GL and PVA/ST blends. The incorporation of HNTs in PVA/ST/GL/HNT bionanocomposite films at the HNT contents of 0.25 and 1 wt% improves both tensile strength and Young's modulus as opposed to those of corresponding polymer blends along with a slight reduction in elongation at break arising from high stiffness of HNTs. These improvements decline beyond 1 wt% HNTs resulting from typical HNT agglomeration, which has been confirmed by an increase in the surface roughness of bionanocomposite films, as well as decreased aspect ratios of embedded HNTs within blend matrices.

Thermal properties of neat PVA in terms of T_g, T_m, ΔH_m and X_c (%) decline with the addition of GL due to GL plasticisation effect. Such properties are increased slightly for PVA/ST blends except X_c (%) owing to amorphous structures of ST. Whereas, PVA/ST/GL blends possess a relatively high T_m when compared with other polymer blends. The presence of HNTs enhances most material properties determined for bionanocomposite films in this chapter, as well as increases the decomposition temperatures relative to those of corresponding blends. Such a finding is associated with the presence of HNTs acting as a barrier material against heat and mass transfer.

The $T\%$ of PVA blends is reduced slightly as opposed to that of neat PVA due to good compatibility of blended polymers. This decreasing tendency becomes more pronounced for bionanocomposite films in the presence of HNTs to increase surface roughness and the number of light scattering sites.

The property improvement for bionanocomposite films is manifested at the HNT content up to 1 wt%, beyond which a slight declining trend has taken place. Such a phenomenon could be explained by minor intercalated structures at the HNT contents of 0.25 to 1 wt% and typical HNT agglomeration beyond 1 wt% accordingly.

References

1. Sorrentino A, Gorrasi G, Vittoria V (2007) Potential perspectives of bio-nanocomposites for food packaging applications. Trend Food Sci Technol 18(2):84–95
2. Rhim JW, Park HM, Ha CS (2013) Bio-nanocomposites for food packaging applications. Prog Polym Sci 38:1629–1652
3. Teodorescu M, Bercea M, Morariu S (2018) Biomaterials of poly(vinyl alcohol) and natural polymers. Polym Rev 58(2):247–287
4. Rhim JW, Ng PKW (2007) Natural biopolymer-based nanocomposite films for packaging applications. Crit Rev Food Sci Nutr 47(4):411–433
5. Mangaraj S, Yadav A, Bal LM, Dash SK, Mahanti NK (2018) Application of biodegradable polymers in food packaging industry: a comprehensive review. J Pack Technol Res 3(1): 77–96
6. Arora A, Padua GW (2010) Review: nanocomposites in food packaging. J Food Sci 75(1):43–49
7. Souza VG, Fernando AL (2016) Nanoparticles in food packaging: biodegradability and potential migration to food—a review. Food Packag Shelf Life 8:63–70
8. Liu JB, Boo WJ, Clearfield A, Sue HJ (2006) Intercalation and exfoliation: a review on morphology of polymer nanocomposites reinforced by inorganic layer structures. Mater Manuf Proc 21(2):143–151
9. Jancar J, Douglas JF, Starr FW, Kumar SK, Cassagnau P, Lesser AJ, Sternstein SS, Buehler MJ (2010) Current issues in research on structure–property relationships in polymer nanocomposites. Polymer 51(15):3321–3343

10. Ismail H, Zaaba NF (2011) Effect of additives on properties of polyvinyl alcohol (PVA)/ tapioca starch biodegradable films. Polym Plast Technol Eng 50(12):1214–1219
11. Cano AI, Cháfer M, Chiralt A, González-Martínez C (2015) Physical and microstructural properties of biodegradable films based on pea starch and PVA. J Food Eng 167:59–64
12. Cano A, Fortunati E, Chafer M, Kenny JM, Chiralt A, Gonzalez-Martínez C (2015) Properties and ageing behaviour of pea starch films as affected by blend with poly(vinyl alcohol). Food Hydrocolloids 48:84–93
13. Wu Z, Wu J, Peng T, Li Y, Lin D, Xing B, Li C, Yang Y, Yang L, Zhang L, Ma R, Wu W, Lv X, Dai J, Han G (2017) Preparation and application of starch/polyvinyl alcohol/citric acid ternary blend antimicrobial functional food packaging films. Polymers 9(3):102
14. Abdullah ZW, Dong Y (2018) Preparation and characterisation of poly(vinyl) alcohol (PVA)/ starch (ST)/halloysite nanotube (HNT) nanocomposite films as renewable materials. J Mater Sci 53(5):3455–3469
15. Khoo WS, Ismail H, Ariffin A (2011) Tensile and swelling properties of polyvinyl alcohol/ chitosan/halloysite nanotubes nanocomposite. Paper presented at the national postgraduate conference. IEEE, Kuala Lumpur, Malaysia, 19–20 Sep 2011
16. Gaaz TS, Sulong AB, Kadhum AAH, Al-Amiery AA, Nassir MH, Jaaz AH (2017) The impact of halloysite on the thermo-mechanical properties of polymer composites. Molecules 22(5): 838 (1–20)
17. Dong Y, Marshall J, Haroosh HJ, Mohammadzadehmoghadam S, Liu D, Qi X, Lau KT (2015) Polylactic acid (PLA)/halloysite nanotube (HNT) composite mats: influence of HNT content and modification. Compos A App Sci Manuf 76:28–36
18. Yuan P, Tan D, Annabi-Bergaya F (2015) Properties and applications of halloysite nanotubes: recent research advances and future prospects. Appl Clay Sci 112–113:75–93
19. Wagner AL, Cooper S, Riedlinger M (2005) Natural nanotubes enhance biodegradable and biocompatible nanocomposites. Ind Biotechnol 1(3):190–193
20. Abdullah ZW, Dong Y, Han N, Liu S (2019) Water and gas barrier properties of polyvinyl alcohol (PVA)/starch (ST)/glycerol (GL)/halloysite nanotube (HNT) bionanocomposite films: experimental characterisation and modelling approach. Compos B Eng 174:107033
21. Farhoodi M, Mousavi SM, Sotudeh-Gharebagh R, Emam-Djomeh Z, Oromiehie A (2014) Migration of aluminum and silicon from PET/Clay nanocomposite bottles into acidic food simulant. Packag Technol Sci 27(2):161–168
22. Voss A, Stark RW, Dietz C (2014) Surface versus volume properties on the nanoscale: elastomeric polypropylene. Macromolecules 47(15):5236–5245
23. Monteiro MKS, Oliveira VRL, Santos FKG, Neto ELB, Leite RHL, Aroucha EMM, Silva RR, Silva KNO (2018) Incorporation of bentonite clay in cassava starch films for the reduction of water vapor permeability. Food Res Int 105:637–644
24. Choudalakis G, Gotsis AD (2009) Permeability of polymer/clay nanocomposites: a review. Eur Polym J 45(4):967–984
25. Sridhar V, Tripathy DK (2006) Barrier properties of chlorobutyl nanoclay composites. J Appl Polym Sci 101(6):3630–3637
26. Saritha A, Joseph K, Thomas S, Muraleekrishnan R (2012) Chlorobutyl rubber nanocomposites as effective gas and VOC barrier materials. Compos A App Sci Manuf 43(6):864–870
27. Azahari NA, Othman N, Ismail H (2011) Biodegradation studies of polyvinyl alcohol/corn starch blend films in solid and solution media. J Phys Sci 22(1):15–31
28. Ramaraj B (2007) Crosslinked poly(vinyl alcohol) and starch composite films: study of their physicomechanical, thermal, and swelling properties. J App PolyM Sci 103(2):1127–1132
29. Sadhu SD, Soni A, Varmani SG, Garg M (2014) Preparation of starch-poly vinyl alcohol (PVA) blend using potato and study of its mechanical properties. Int J Pharmaceut Sci Invent 3(3):33–37
30. Tang S, Zou P, Xiong H, Tang H (2008) Effect of nano-SiO$_2$ on the performance of starch/ polyvinyl alcohol blend films. Carbohydr Polym 72(3):521–526

31. Qiu K, Netravali AN (2013) Halloysite nanotube reinforced biodegradable nanocomposites using noncrosslinked and malonic acid crosslinked polyvinyl alcohol. Polym Compos 34 (5):799–809
32. Heidarian P, Behzad T, Sadeghi M (2017) Investigation of cross-linked PVA/starch biocomposites reinforced by cellulose nanofibrils isolated from aspen wood sawdust. Cellulose 24(8):3323–3339
33. Cano A, Fortunati E, Chafer M, Gonzalez-Martınez C, Chiralt A, Kenny JM (2015) Effect of cellulose nanocrystals on the properties of pea starch–poly(vinyl alcohol) blend films. J Mater Sci 50(21):6979–6992
34. Mohsin M, Hossin A, Haik Y (2011) Thermomechanical properties of poly(vinyl alcohol) plasticized with varying ratios of sorbitol. Mater Sci Eng A 528:925–930
35. Aydın AA, Ilberg V (2016) Effect of different polyol-based plasticizers on thermal properties of polyvinyl alcohol: starch blends. Carbohydr Polym 136:441–448
36. Liu M, Jia Z, Jia D, Zhou C (2014) Recent advance in research on halloysite nanotubes-polymer nanocomposite. Prog Polym Sci 39(8):1498–1525
37. Nistor MT, Vasile C (2012) Influence of the nanoparticle type on the thermal decomposition of the green starch/ poly(vinyl alcohol)/ montmorillonite nanocomposites. J Thermal Analy Calorim 111(3):1903–1919
38. Sin LT, Rahman WAWA, Rahmat AR, Mokhtar M (2011) Determination of thermal stability and activation energy of polyvinyl alcohol–cassava starch blends. Carbohydr Polym 83 (1):303–305
39. Sreekumar PA, Al-Harthi MA, De SK (2012) Effect of glycerol on thermal and mechanical properties of polyvinyl alcohol/starch blends. J App Polym Sci 123:135–142
40. Nistor MT, Vasile C (2013) TG/FTIR/MS study on the influence of nanoparticles content upon the thermal decomposition of starch/poly(vinyl alcohol) montmorillonite nanocomposites. Iranian Polym J 22(7):519–536
41. Priya B, Gupta VK, Pathania D, Singha AS (2014) Synthesis, characterization and antibacterial activity of biodegradable starch/PVA composite films reinforced with cellulosic fibre. Carbohydr Polym 109:171–179
42. Sadhu SD, Soni A, Garg M (2015) Thermal studies of the starch and polyvinyl alcohol based film and its nano composites. J Nanomedic Nanotechnol S7:002
43. Lim M, Kwon H, Kim D, Seo J, Han H, Khan SB (2015) Highly-enhanced water resistant and oxygen barrier properties of cross-linked poly(vinyl alcohol) hybrid films for packaging applications. Prog Organic Coating 85:68–75
44. Sreedhar B, Sairam M, Chattopadhyay DK, Rathnam PAS, Rao DVM (2005) Thermal, mechanical, and surface characterization of starch-poly(vinyl alcohol) blends and borax-crosslinked films. J Appl Polym Sci 96(4):1313–1322
45. Strawhecker KE, Manias E (2000) Structure and properties of poly(vinyl alcohol)/Na⁺ montmorillonite nanocomposites. Chemist Mater 12:2943–2949
46. Jose J, Al-Harthi MA, Al-Ma'adeed MA, Dakua JB, De SK (2015) Effect of graphene loading on thermomechanical properties of poly(vinyl alcohol)/starch blend. J App Polym Science 132(16):41827
47. Liu M, Guo B, Du M, Jia D (2007) Drying induced aggregation of halloysite nanotubes in polyvinyl alcohol/halloysite nanotubes solution and its effect on properties of composite film. Appl Phys A 88(2):391–395
48. Tee TT, Sin LT, Gobinath R, Bee ST, Hui D, Rahmat AR, Kong I, Fang QH (2013) Investigation of nano-size montmorillonite on enhancing polyvinyl alcohol–starch blends prepared via solution cast approach. Compos B Eng 47:238–247
49. Dong Y, Bickford T, Haroosh HJ, Lau KT, Takagi H (2013) Multi-response analysis in the material characterisation of electrospun poly (lactic acid)/halloysite nanotube composite fibres based on Taguchi design of experiments: fibre diameter, non-intercalation and nucleation effects. Appl Phys A 112:747–757
50. Akhavan A, Khoylou F, Ataeivarjovi E (2017) Preparation and characterization of gamma irradiated Starch/PVA/ZnO nanocomposite films. Radia Phys Chemist 138:49–53

51. Gupta B, Agarwal R, Alam MS (2013) Preparation and characterization of polyvinyl alcohol-polyethylene oxide-carboxymethyl cellulose blend membranes. J App Polym Sci 127 (2):1301–1308
52. Guohua Z, Ya L, Cuilan F, Min Z, Caiqiong Z, Zongdao C (2006) Water resistance, mechanical properties and biodegradability of methylated-cornstarch/poly(vinyl alcohol) blend film. Polym Degrad Stab 91(4):703–711
53. Abdullah ZW, Dong Y (2019) Biodegradable and water resistant poly(vinyl) alcohol (PVA)/ starch (ST)/glycerol (GL)/halloysite nanotube (HNT) nanocomposite films for sustainable food packaging. Frontiers Mater 6:58
54. Grunlan JC, Grigorian A, Hamilton CB, Mehrabi AR (2004) Effect of clay concentration on the oxygen permeability and optical properties of a modified poly(vinyl alcohol). J Appl Polym Sci 93(3):1102–1109
55. Lee MH, Kim SY, Park HJ (2018) Effect of halloysite nanoclay on the physical, mechanical, and antioxidant properties of chitosan films incorporated with clove essential oil. Food Hydrocoll 84:58–67

Chapter 4
Water Resistance and Biodegradability of PVA/HNT Bionanocomposite Films

Abstract Water resistance is evaluated for polyvinyl alcohol (PVA)/starch (ST)/ glycerol (GL)/halloysite nanotube (HNT) bionanocomposite films as one of the fundamental requirements for food packaging materials. The incorporation of HNTs at the HNT contents between 0.25 and 1 wt% reduces the water absorption capacity and water solubility of bionanocomposite films by 42 and 48.05%, respectively, when compared with those of PVA/ST/GL blends. This is because of the moderate hydrophobicity of HNTs, as evidenced by increasing water contact angle of bionanocomposite films linearly by 21.36° with increasing the HNT content from 0 to 5 wt%, while slight improvements in water absorption capacity and water solubility of bionanocomposite films have been recorded beyond 1 wt% HNTs due to their typical agglomeration issue. Moreover, the presence of HNTs reduces the biodegradation rate of bionanocomposite films as opposed to PVA/ST/ GL blends despite still being better than neat PVA films in view of their morphological structures.

Keywords Hydrophilic polymers · Water resistance · Biodegradation · Morphological structure · Fungal hyphae

In general, water-soluble polymers have poor water resistance, as reflected by high water uptake and high water solubility due to their free hydroxyl groups that can be easily interacted with water molecules [1]. Several methods have been used to improve the water resistance of these polymers such as blending with other polymers [2], coating [3], using cross-linking agents [4], ionising rays like ultraviolet rays (UV) [5] and nanotechnology [6]. All these methods based on consuming free hydroxyl groups and/or increased the surface hydrophobicity of polymers [4]. In this chapter, HNTs as moderately hydrophobic nanofillers are used to improve the water resistance of PVA/ST/GL blends. On the other hand, ST is generally considered as a good material candidate to enhance the biodegradability of PVA with high cost reduction [7, 8]. Consequently, the effect of ST on the biodegradation rates of PVA blends and bionanocomposites is investigated in detail as well.

4.1 Water Absorption Capacity (W_a)

Water absorption capacity or water uptake (W_a) is an important parameter to evaluate bionanocomposite materials particularly targeting food packaging applications [9, 10]. W_a% of neat PVA, PVA blends and PVA/ST/GL/HNT bionanocomposite films are illustrated in Fig. 4.1. As compared with neat PVA, PVA/GL blends have a slight decrease in W_a by 10.21% due to good compatibility between PVA and GL leading to the consumption of more free hydroxyl groups, which is in good agreement with Follain et al. [11]. Whereas, PVA/ST blends have the highest W_a owing to partial compatibility between PVA and ST in the absence of plasticisers leading to more free sites in the blends to be occupied by water molecules. Additionally, ST has high hydrophilicity resulting in increasing the W_a of ST blends, which is associated with the hygroscopic nature of ST for the water gain or loss in order to achieve the equilibrium with the environment [12–15]. Nonetheless, the presence of GL reduces the W_a of PVA/ST/GL blends by 23.92% as opposed to that of PVA/ST blend counterparts, which is ascribed to the improvement of compatibility and interactions between blend constituents. Similarly, Zou et al. [16] stated that the addition of GL reduced the W_a of PVA/ST blends by enhancing the compatibility between components.

A further decrease in W_a has been achieved with the incorporation of HNTs in bionanocomposite films due to the typical restriction of water diffusion in the presence of HNTs resulting from the tortuous paths, as opposed to that of PVA/ST/

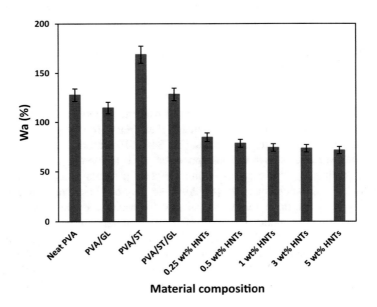

Fig. 4.1 Water absorption capacity of neat PVA, PVA blends and PVA/ST/GL/HNT bionanocomposite films at different HNT contents [17]

GL blends. A remarkable reduction of W_a by 42.0% has been gained for PVA/ST/ GL/HNT bionanocomposites with increasing the HNT content from 0.25 to 1 wt% on account of good dispersion of nanofillers with the restriction to the diffusion of water molecules. Lee et al. [18] reported similar effect of HNTs on the W_a of chitosan/HNT nanocomposites and chitosan/clove essential oil (CEO)/HNT nanocomposites relative to their corresponding biopolymers or blends. The W_a of chitosan/HNT nanocomposites and chitosan/CEO/HNT nanocomposites declined by 42.17 and 43.31%, respectively with the addition of 30 wt% HNTs, as opposed to those of their matrices alone due to the presence of moderately hydrophobic nanofillers. Moreover, Abbasi [19] concluded that the W_a of PVA/ST/SiO$_2$ nanocomposites declined by 50% with increasing the SiO$_2$ nanofiller content from 1 to 5 wt%, which arose from strong physical interactions between components in order to consume more free hydroxyl groups that could bond with water molecules. Additionally, a slight reduction in W_a by 4.04% has been observed for bionanocomposites at the HNT contents beyond 1 wt% owing to typical HNT agglomeration at 3 and 5 wt%, as previously evidenced in morphological observations via SEM and AFM in Chap. 3. These findings are consistent with other results based on PVA/ST/MMT nanocomposites [20, 21], PVA/ST/nano-SiO$_2$ nanocomposites [22] and PVA/ST/ZnO nanocomposites [23].

When compared with other nanofillers, HNTs produce a higher reduction in W_a at low nanofiller contents up to 1 wt% due to their moderate hydrophobicity, as shown in Fig. 4.2. Whereas, this reduction becomes less pronounced beyond 1 wt% as compared with other nanofillers with a similar behaviour at different nanofiller contents because of nanofiller agglomeration. Overall, nanofiller content and dispersion are considered as major factors to affect the reduction in the W_a of nanocomposite films.

4.2 Water Solubility (W_s)

Water solubility (W_s) is one of critical material characteristics in relation to water resistance, especially for water-soluble polymers like PVA. According to Azahari et al. [14], when the material has high water absorption capacity, it possesses high water solubility as well because water molecules are absorbed onto hydroxyl groups particularly on hydrogen bonding sites leading to weak material structures and easier water dissolution. As such, W_a and W_s of neat PVA, PVA blends and PVA/ ST/GL/HNT bionanocomposite films at different HNT contents show a very similar trend despite their different magnitudes, as demonstrated in Fig. 4.3. These films have a relatively high W_s because of high hydrophilicity of all constituents like PVA, ST and GL, in good accordance with other studies [24, 25]. PVA/GL blends yield a slight decrease in the W_s by 8.21% as opposed to that of neat PVA films because the number of free hydroxyl groups could decrease in polymer blends with the addition of GL. However, the W_s of PVA/ST blends increases by 4.69% when compared with that of neat PVA due to the increased hydrophilicity of blends with

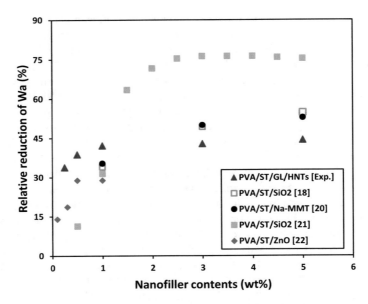

Fig. 4.2 Data comparison for the relative reduction in W_a% of PVA/ST nanocomposite films reinforced with different types of nanofillers

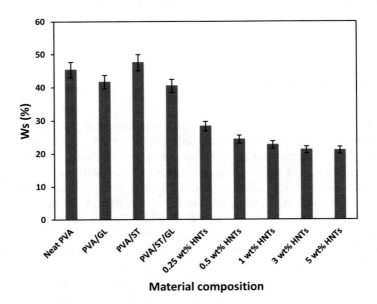

Fig. 4.3 Water solubility of neat PVA, PVA blends and PVA/ST/GL/HNT bionanocomposite films at different HNT contents [17]

the addition of ST, as confirmed by Negim et al. [26]. Then, the W_s of PVA/ST/GL blends is reduced by 14.89% as opposed to that of PVA/ST films with additional GL leading to the enhancement of compatibility and interfacial bonding between PVA and ST [24, 25].

When compared with corresponding polymer blends alone, PVA/ST/GL/HNT bionanocomposite films possess a remarkable reduction in W_s by 48.05% with increasing the HNT content from 0.25 to 5 wt%. This reduction may be interpreted by the moderate hydrophobicity of HNTs resulting from the low number of hydroxyl groups distributed on HNT surfaces [27]. Additionally, the SiO_2 groups of HNTs have the ability to form numerous hydrogen bonds in nanocomposite films so as to restrict mass transfer, as well as consume the hydroxyl groups of polymer matrices leading to decreasing the number of free interaction sites occupied by water molecules [9]. The good dispersion of HNTs up to 1 wt% gives rise to a remarkable reduction in W_s of bionanocomposite films when compared with those films beyond 1 wt% HNTs with minor reduction only by 7.2%. This finding can be associated with HNT agglomeration mentioned in Sect. 4.1. Similar results are also manifested based on PVA/ST/nano-SiO_2 nanocomposites [22], PVA/ST/$CaCO_3$ nanocomposites [28] and PVA/ST/nanotitania nanocomposites [29].

4.3 Water Contact Angle

The water contact angles of neat PVA, PVA blends and PVA/ST/GL/HNT bio-nanocomposite at different HNT contents are evaluated to understand the hydrophilicity of material surfaces, as illustrated in Fig. 4.4. It is well known that water contact angle less than 90° means high material wettability while it is referred to as low wettability when greater than 90° instead [30]. In other words, the materials with low water contact angle may be associated with the high hydrophilicity of their surfaces and vice versa [9, 31]. Neat PVA as a water-soluble polymer has a low water contact angle of 28.35° measured in this work in good accordance with previous studies [32]. In contrast with neat PVA, PVA/GL blends have slightly higher water contact angle by 0.33°, which is primarily associated with the insignificant reduction in W_a and W_s for PVA/GL blends. Moreover, the water contact angle of PVA/ST blends decreases by 9.7° as opposed to that of neat PVA since the addition of ST can improve the hydrophilicity. As such, a reduction of water contact angle by 2.78° has been found for PVA/ST/GL blends relative to that of neat PVA films. As clearly shown in Fig. 4.5, the water contact angles of PVA/ST and PVA/ST/GL blends are remarkably increased leading to the improvement of hydrophilicity in contrast with that of neat PVA thanks to more hydroxyl groups in PVA blends in the presence of ST. These findings can explain the further increases in W_a and W_s of such blends.

A linear increasing trend in water contact angle of bionanocomposite films has been achieved from 25.57° to 46.93° with increasing the HNT content from 0 to 5 wt%. HNTs as moderately hydrophobic nanofillers have a low number of

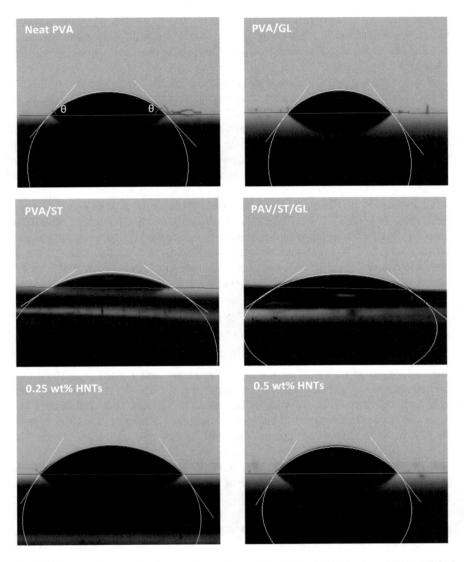

Fig. 4.4 Images of water droplets on the surfaces of neat PVA, PVA blends and PVA/ST/GL/ HNT bionanocomposite films for contact angle measurements

hydroxyl groups to play an important role in the reduction in W_a and surface hydrophilicity [9, 27, 31]. Additionally, well-dispersed HNTs within polymer matrices enable to consume some free hydroxyl groups in nanocomposites in order to generate hydrogen bonding between components. It is well known that the incorporation of nanofillers can increase surface roughness of nanocomposites [33]. Therefore, increasing surface roughness of nanocomposites in the presence of

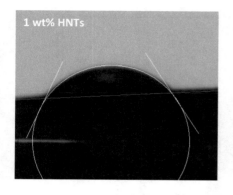
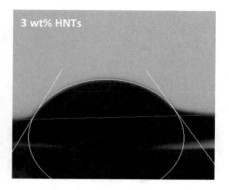
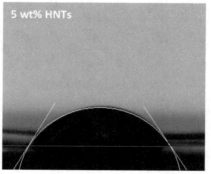

Fig. 4.4 (continued)

nanofillers can lead to a further increase in water contact angle according to Wenzel's theory [34]. As such, the surface hydrophobicity is increased with increasing surface roughness based on the equation as follows [35]:

$$\cos \theta_m = r \cos \theta \tag{4.1}$$

where θ_m and θ are the measured and ideal water contact angles, respectively, which can be calculated from the perfect smooth surface like mirror. r is the surface roughness ratio where $r = 1$ for smooth surfaces and $r > 1$ for rough surfaces [35]. Similarly, the water contact angle of pectin/HNT nanocomposite film surfaces has been shown to be increased by 6° as compared with that of pectin counterparts due to increased surface roughness of nanocomposites in the presence of HNTs [36].

Although the water contact angles of PVA/ST/GL/HNT bionanocomposite films are increased by 21.36° compared with that of corresponding blends counterparts, the films could still be categorised within the range of hydrophilic materials (i.e. $\theta < 90°$). Overall, the moderate hydrophobicity of HNTs is the main reason for increasing water contact angles on the surfaces of bionanocomposite films.

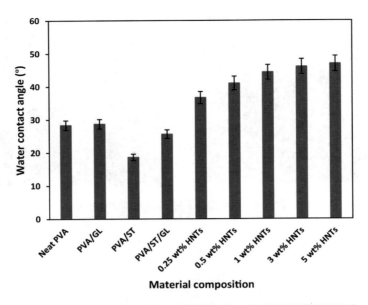

Fig. 4.5 Water contact angles of neat PVA, PVA blends and PVA/ST/GL/HNT bionanocomposite films at different HNT contents [17]

4.4   Soil Burial Biodegradation

In general, the biodegradability of most biopolymers depends on their W_a and W_s because the degradation process starts with water absorption on the material surfaces, which is then followed by the growth of microorganisms like bacteria and fungi to finish with cleavage particularly for soil burial biodegradation [24, 37]. The digital images of neat PVA, PVA blend and PVA/ST/GL/HNT bionanocomposite samples before and after soil burial biodegradation tests over the period of 24 weeks are presented in Fig. 4.6. Relative to neat PVA films, all other material films diminish in size after the test period and change to more fragile and wrinkling films. Moreover, neat PVA and PVA/GL blend films still have good transparency after biodegradation tests, while other material films show great colour change and tend to be more yellowish rather than transparent after biodegradation tests. These changes are greatly related to the presence of potato-based ST prone to the attack first by microorganisms as completely biodegradable materials when compared with PVA counterparts [38].

Biodegradation rates of neat PVA, PVA blend and PVA/ST/GL/HNT bionanocomposite films at different HNT contents have been determined as a function of time, as demonstrated in Fig. 4.7. The biodegradation rates of all material films over 24 weeks can be characterised over an active-state period in the first three weeks where the material samples degrade at a very rapid rate. This is followed by a steady-state period for the rest of time when the material samples degrade at a

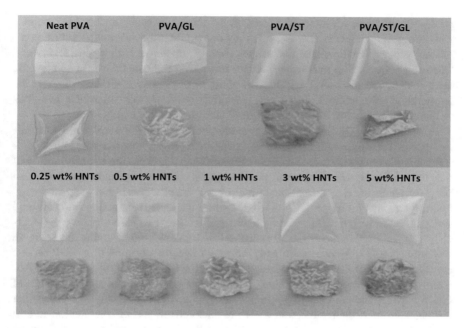

Fig. 4.6 Digital images of neat PVA, PVA blend and PVA/ST/GL/HNT bionanocomposite films before (0 week) and after (24 weeks) in soil burial biodegradation tests [17]

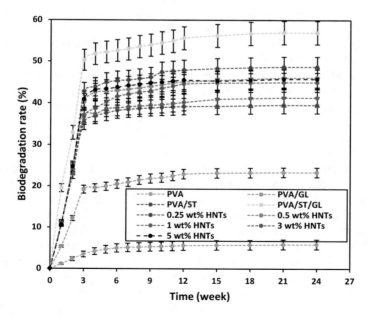

Fig. 4.7 Biodegradation rates of neat PVA, PVA blend and PVA/ST/GL/HNT bionanocomposite films at different HNT contents [17]

relatively slow rate till the end of tests, as evidenced by other studies [37, 39, 40]. According to Azahari et al. [14], this phenomenon can be associated with a composting process consisting of two stages, namely "an active composting stage" and "a curing period". The first stage includes increasing the temperature at an elevated level due to the microbial activities as long as oxygen is available. Further, the temperature decreases in stage two resulting from slower microbial activities despite continued slow degradation rates.

Neat PVA has the lowest biodegradation rates of 5.87% among all materials because of its high resistance to biodegradation in soil compared with other environments like sludge, while the slight weight loss is related to the hydrolysability of neat PVA [37, 41, 42]. In general, polymers with carbon-only backbones like most vinyl polymers are not susceptible to hydrolysis and biodegradation [43]. However, hydroxyl groups (–OH) of PVA are oxidised by the enzymatic action into carbonyl groups (C=O), which is followed by the hydrolysis into two carbonyl groups (–CO–CH_2–CO–). This induces the cleavage of polymeric chains and a decrease in molecular weight. Hence, such low molecular weight portions of neat PVA are consumed by microbes [43]. On the other hand, the biodegradation rate of PVA/GL blends is increased up to 23.33% relative to that of neat PVA due to the presence of GL to improve the mobility of polymeric chains. This results in increasing the water diffusion through their molecular structures. Moreover, the biodegradation rate of PVA/ST blends is increased further up to 39.54% due to the addition of ST, which is regarded as a fully biodegradable polymer [14, 37] whose material structures can be easily attacked by microorganisms [39, 43]. The biodegradation rates of PVA/ST blends increase linearly with increasing the ST content, as reported in previous work [14, 38, 44]. Further improvement of biodegradation rates up to 56.94% is demonstrated in PVA/ST/GL blends owing to the combination of complete biodegradation of ST with GL plasticisation effect to improve microorganism infiltration rate. As such, PVA/ST/GL blends have a biodegradation rate comparable to those of other natural biopolymers shown in Fig. 4.8.

The biodegradability of PVA/ST/GL/HNT bionanocomposite films decline significantly from 56.94 to 41.28% with increasing the HNT content from 0 to 1 wt %. This phenomenon is associated with good dispersion of HNTs within polymer blend matrices leading to strong hydrogen bonds between them to hinder water diffusion, mass transfer and microorganism infiltration rate [22]. Biodegradation rates of bionanocomposite films tend to increase slightly up to 45 and 45.8% at the HNT contents of 3 and 5 wt%, respectively. Such results are attributed to HNT agglomeration even though these rates are still higher than that of neat PVA counterparts, as evidently shown in other studies [22, 41, 46].

Well-dispersed HNTs clearly reduce the biodegradation rates of bionanocomposite films as opposed to other nanofillers at the same contents up to 1 wt% due to the strong interfacial bonding between HNTs and blend matrice. Whereas, HNT agglomeration happening at the HNT contents of 3 and 5 wt% in bionanocomposite films diminishes these interactions and their effect on biodegradation rates, as illustrated in Fig. 4.9.

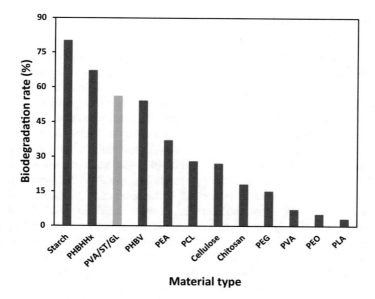

Fig. 4.8 Comparisons for biodegradation rates of PVA/ST/GL blends with different other material types with the data collected from [45] with copyright permission from Springer. *PHBHHx* poly(3-hydroxybutyrate-co-3-hydroxyhexanoate), *PHBV* (3-hydroxybutyrate- co - 3-hydroxyvalerate), *PEA* polyethylene acrylate, *PCL* polycaprolactone, *PEG* polyethylene glycol, *PEO* polyethylene oxide and *PLA* polylactic acid

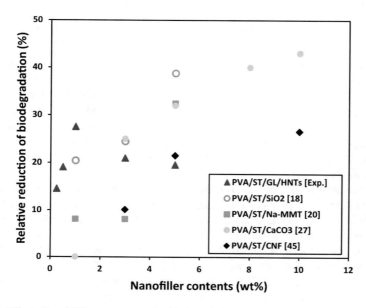

Fig. 4.9 Effect of nanofiller content on the biodegradation rates of PVA/ST bionanocomposite films reinforced with different nanofillers

4.5   SEM Observation

In order to understand different morphological structures taking place in soil burial degradation tests, neat PVA, PAV blends and PVA/ST/GL/HNT bionanocomposite films at various HNT contents were examined via SEM. Moreover, morphological structures of these samples were evaluated before the biodegradation tests in the initial week and during the active stage period after 1 and 3 weeks, as well as at the end of biodegradation tests after 24 weeks according to Table 4.1.

At an initial week, both neat PVA and PVA/GL blend films show similar smooth surface morphology in good accordance with [47, 48]. Such smooth structures are completely replaced with rough globular morphology for PVA/ST blends due to the partial compatibility between PVA and ST in the absence of plasticisers [48]. As a result, two separate PVA-rich (continuous phase) and ST-rich (globular structure) phases have been clearly observed. This partial compatibility reported in PVA/ST blends is considered as the main reason for their higher W_a and W_s in Sects. 4.1 and 4.2, respectively. Smooth surface morphology appears again in PVA/ST/GL blends as a result of the addition of GL to promote the compatibility between components and meanwhile slightly reduce W_a and W_s when compared with those of PVA/ST blends. The addition of HNTs does not show phase separation of bionanocomposite films though the surface roughness is increased particularly with the inclusion of 3 and 5 wt% HNTs, which is in good agreement with the AFM results obtained in Chap. 3.

After 1 week of soil burial biodegradation tests, neat PVA films do not show clear surface changes, which is confirmed by low biodegradation rates in Sect. 4.4, while branched traces known as "fungal hyphae" [49] are slightly observed on PVA/GL blend surfaces, and become more clearly presented on surfaces of PVA/ST and PVA/ST/GL blend films. According to Qiu and Netravali [49], these fungal hyphae are commonly generated during the progress in biodegradation process at appropriate temperature levels. Moreover, the number and extension of these fungal hyphae increases with time to cross each other and form grooves [50]. Additionally, the presence of fungal hyphae is good evidence of high biodegradation rates during an active stage period of PVA blend films. Fungal hyphae do not appear on bio-nanocomposite films despite the increase in surface roughness because of the presence of HNTs. This reduces microorganism infiltration rate, as evidenced by Tang et al. [22].

After 3 weeks, the surface roughness of neat PVA films is increased without sensible changes, while clear alterations are observed on PVA/GL blend films like the pore formation with average diameters in range of 0.12–1.64 μm. The addition of GL highly improves the mobility of polymeric chains [13, 51] leading to increasing microorganism infiltration rate and biodegradation rate of PVA/GL blends. Fungal hyphae are connected to each other in PVA/ST and PVA/ST/GL blend films to generate small open pores with much faster biodegradation rates as opposed to that of neat PVA film counterparts. When compared with PVA/ST/GL

Table 4.1 SEM micrographs of neat PVA, PVA blends and PVA/ST/GL/HNT bionanocomposites at different HNT contents at initial week, after 1 week, after 3 weeks and 24 weeks in soil burial degradation tests [17]

Material	Initial week	After 1 week	After 3 weeks	After 24 weeks
Neat PVA				
PVA/GL				

(continued)

Table 4.1 (continued)

Material	Initial week	After 1 week	After 3 weeks	After 24 weeks
PVA/ST				
PVA/ST/GL				

(continued)

Table 4.1 (continued)

Material	Initial week	After 1 week	After 3 weeks	After 24 weeks
0.25 wt% HNTs				
0.5 wt% HNTs				

(continued)

Table 4.1 (continued)

Material	Initial week	After 1 week	After 3 weeks	After 24 weeks
1 wt% HNTs				
3 wt% HNTs				

(continued)

Table 4.1 (continued)

Material	Initial week	After 1 week	After 3 weeks	After 24 weeks
5 wt% HNTs				

blends alone, bionanocomposite films possess a small number of pores with relatively large average diameters in range of 0.21–3.05 µm.

After 24 weeks of soil burial biodegradation tests, tiny pores with relatively small average diameters of 0.06–0.72 µm are revealed with more or less uniform distribution on the surfaces of neat PVA films, which is reflected by the low biodegradation rate of neat PVA. However, the progressive deterioration from the external to internal layers is revealed in PVA blends and PVA/ST/GL/HNT bionanocomposite films. Overall, the addition of ST has remarkable effects on the biodegradation rates and morphological structures of PVA/ST blends, PVA/ST/GL blends and bionanocomposite films relative to those of neat PVA and PVA/GL blends.

4.6   Summary

Neat PVA and ST are well known as hydrophilic polymers, so their blends and bionanocomposites have relatively high W_a and W_s. When compared with neat PVA, PVA/GL blends have slightly lower W_a and W_s due to the consumption of some free hydroxyl groups by GL molecules to produce hydrogen bonds. A completely opposite trend is shown for PVA/ST blends because the partial compatibility between them makes many free hydroxyl groups occupied by water molecules. This compatibility is promoted slightly with the addition of GL leading to the reduction in the W_a and W_s of PVA/ST/GL blends. These findings have been further confirmed by the decreased water contact angles of PVA blend films as compared with that of neat PVA films.

A remarkable reduction in W_a and W_s has been reported with the incorporation of HNTs within bionanocomposite films owing to moderate hydrophobicity of HNTs, as well as the formation of strong hydrogen bonds between nanofillers and blend matrices to restrict the diffusion of water molecules and mass transfer. The presence of HNTs significantly increases water contact angles of bionanocomposite films as a result of improving their hydrophobicity. The HNT agglomeration at the content levels of 3 and 5 wt% can restrict the reduction in the W_a and W_s of bionanocomposite films despite being still better than those of PVA/ST/GL blends.

The poor biodegradability of PVA in some environments like soil is the main reason for lower biodegradation rate and slight morphological changes. This rate is increased for PVA/GL blends thanks to the enhanced mobility of polymeric chains to increase microorganism infiltration rate. Such an increasing tendency becomes more pronounced for PVA/ST blends with the addition of ST as a completely biodegradable polymer. Moreover, the combination of ST and GL highly promotes the biodegradation rate of PVA/ST/GL blends. This phenomenon is reflected by the growth of fungal hyphae over an active stage period and apparent damages at the steady-state period.

The good dispersion of HNTs reduces the biodegradation rates of PVA/ST/GL/ HNT bionanocomposite films in a linear manner with increasing the HNT content from 0.25 to 1 wt% because hydrogen bonding networks are established between nanofillers and blend matrices, thus reducing mass transfer. The biodegradation rate of bionanocomposite films is slightly increased beyond 1 wt% HNTs as a result of typical HNT agglomeration, which destabilises the molecular bonding between HNTs and blend matrices.

References

1. Chiellini E, Corti A, D'Antone S, Solaro R (2003) Biodegradation of poly (vinyl alcohol) based materials. Prog Polym Sci 28:963–1014
2. Gupta B, Agarwal R, Alam MS (2013) Preparation and characterization of polyvinyl alcohol-polyethylene oxide-carboxymethyl cellulose blend membranes. J App Polym Sci 127 (2):1301–1308
3. Khwaldia K, Arab-Tehrany E, Desobry S (2010) Biopolymer coatings on paper packaging materials. CRFSFS 9:82–91
4. Maiti S, Ray D, Mitra D (2012) Role of crosslinker on the biodegradation behavior of starch/polyvinylalcohol blend films. J Polym Environ 20:749–759
5. Shahabi-Ghahfarrokhi I, Goudarzi V, Babaei-Ghazvini A (2019) Production of starch based biopolymer by green photochemical reaction at different UV region as a food packaging material: physicochemical characterization. Int J Biol Macromol 122:201–209
6. Abdollahi M, Alboofetileh M, Behrooz R, Rezaei M, Reza Miraki R (2013) Reducing water sensitivity of alginate bio-nanocomposite film using cellulose nanoparticles. Int J Biol Macromol 54:166–173
7. Jose J, Al-Harthi MA, Al-Ma'adeed MA, Dakua JB, De SK (2015) Effect of graphene loading on thermomechanical properties of poly(vinyl alcohol)/starch blend. J App Polym Science 132(16):41827
8. Sadhu SD, Soni A, Garg M (2015) Thermal studies of the starch and polyvinyl alcohol based film and its nano composites. J Nanomedic Nanotechnol S7:002
9. Sadegh-Hassani F, Nafchi AM (2014) Preparation and characterization of bionanocomposite films based on potato starch/halloysite nanoclay. Int J Biol Macromol 67:458–462
10. Aloui H, Khwaldia K, Hamdi M, Fortunati E, Kenny JM, Buonocore GG, Lavorgna M (2016) Synergistic effect of halloysite and cellulose nanocrystals on the functional properties of PVA based nanocomposites. ACS Sust Chemist Eng 4(3):794–800
11. Follain N, Joly C, Dole P, Bliard C (2005) Properties of starch based blends. Part 2. Influence of poly vinyl alcohol addition and photocrosslinking on starch based materials mechanical properties. Carbonhydr Polym 60(2):185–192
12. Ali M (2016) Synthesis and study the effect of HNTs on PVA/chitosan composite material. Int J Chem Molecul Nucl Mater Metallur Eng 10:234–240
13. Ismail H, Zaaba NF (2011) Effect of additives on properties of polyvinyl alcohol (PVA)/tapioca starch biodegradable films. Polym Plast Technol Eng 50(12):1214–1219
14. Azahari NA, Othman N, Ismail H (2011) Biodegradation studies of polyvinyl alcohol/corn starch blend films in solid and solution media. J Phys Sci 22(2):15–31
15. Salleh MSN, Mohamed Nor NN, Mohd N, Syed Draman SF (2017) Water resistance and thermal properties of polyvinyl alcohol-starch fiber blend film. Paper presented at AIP conference proceedings 1809, American Institute of Physics, Feb 2017
16. Zou GX, Ping-Qu J, Liang-Zou X (2008) Extruded starch/PVA composites: water resistance, thermal properties, and morphology. J Elastom Plast 40(4):303–316
17. Abdullah ZW, Dong Y (2019) Biodegradable and water resistant poly(vinyl) alcohol (PVA)/starch (ST)/glycerol (GL)/halloysite nanotube (HNT) nanocomposite films for sustainable food packaging. Frontiers Mater 6:58

18. Lee MH, Kim SY, Park HJ (2018) Effect of halloysite nanoclay on the physical, mechanical, and antioxidant properties of chitosan films incorporated with clove essential oil. Food Hydrocoll 84:58–67
19. Abbasi Z (2012) Water resistance, weight loss and enzymatic degradation of blends starch/polyvinyl alcohol containing SiO_2 nanoparticle. J Taiwan Instit Chem Eng 43:264–268
20. Tian H, Wang K, Liu D, Yan J, Xiang A, Rajulu AV (2017) Enhanced mechanical and thermal properties of poly (vinyl alcohol)/corn starch blends by nanoclay intercalation. Int J Biol Macromol 101:314–320
21. Taghizadeh MT, Abbasi Z, Nasrollahzade Z (2011) Study of enzymatic degradation and water absorption of nanocomposites starch/polyvinyl alcohol and sodium montmorillonite clay. J Taiwan Instit Chem Eng 43:120–124
22. Tang S, Zou P, Xiong H, Tang H (2008) Effect of nano-SiO_2 on the performance of starch/polyvinyl alcohol blend films. Carbohydr Polym 72(3):521–526
23. Akhavan A, Khoylou F, Ataeivarjovi E (2017) Preparation and characterization of gamma irradiated Starch/PVA/ZnO nanocomposite films. Rad Phys Chem 138:49–53
24. Zanela J, Olivato JB, Dias AP, Grossmann MVE, Yamashita F (2015) Mixture design applied for the development of films based on starch, polyvinyl alcohol, and glycerol. J Appl Polym Sci 132(43):42697
25. Cano AI, Cháfer M, Chiralt A, González-Martínez C (2015) Physical and microstructural properties of biodegradable films based on pea starch and PVA. J Food Eng 167:59–64
26. Negim ESM, Rakhmetullayeva RK, Yeligbayeva GZh, Urkimbaeva PI, Primzharova ST, Kaldybekov DB, Khatib JM, Mun GA, Craig W (2014) Improving biodegradability of polyvinyl alcohol/starch blend films for packaging applications. Int J Basic Appl Sci 3(3): 263–273
27. Liu M, Jia Z, Jia D, Zhou C (2014) Recent advance in research on halloysite nanotubes-polymer nanocomposite. Prog Polym Sci 39(8):1498–1525
28. Kisku SK, Sarkar N, Dash S, Swain SK (2014) Preparation of starch/PVA/$CaCO_3$ nanobiocomposite films: study of fire retardant, thermal resistant, gas barrier and biodegradable properties. Polym Plast Technol Eng 53:1664–1670
29. Lin D, Huang Y, Liu Y, Luo T, Xing B, Yang Y, Yang Z, Wu Z, Chen H, Zhang Q, Qin W (2018) Physico-mechanical and structural characteristics of starch/polyvinyl alcohol/nano-titania photocatalytic antimicrobial composite films. LWT Food Sci Technol 96:704–712
30. Yuan Y, Lee TR (2013) Contact Angle and Wetting Properties. In: Bracco G, Holst B (eds) Surface science techniques. Springer, Berlin, pp 3–34
31. Alipoormazandarani N, Ghazihoseini S, Nafchi AM (2015) Preparation and characterization of novel bionanocomposite based on soluble soybean polysaccharide and halloysite nanoclay. Carbohyd Polym 134:745–751
32. Lim M, Kwon H, Kim D, Seo J, Han H, Khan SB (2015) Highly-enhanced water resistant and oxygen barrier properties of cross-linked poly(vinyl alcohol) hybrid films for packaging applications. Prog Organic Coating 85:68–75
33. Grunlan JC, Grigorian A, Hamilton CB, Mehrabi AR (2004) Effect of clay concentration on the oxygen permeability and optical properties of a modified poly(vinyl alcohol). J Appl Polym Sci 93(3):1102–1109
34. Wenzel RN (1949) Surface roughness and contact angle. J Phys Colloid Chemist 53(9): 1466–1467
35. Kubiak KJ, Wilson MCT, Mathia TG, Carval PH (2011) Wettability versus roughness of engineering surfaces. Wear 271:523–528
36. Biddeci G, Cavallaro G, Blasi FD, Lazzara G, Massaro M, Milioto S, Parisi F, Riela S, Spinelli G (2016) Halloysite nanotubes loaded with peppermint essential oil as filler for functional biopolymer film. Carbohydr Polym 152:548–557
37. Guohua Z, Ya L, Cuilan F, Min Z, Caiqiong Z, Zongdao C (2006) Water resistance, mechanical properties and biodegradability of methylated-cornstarch/poly(vinyl alcohol) blend film. Polym Degrad Stab 91(4):703–711

38. Tănase EE, Popa ME, Rapa M, Popa O (2015) Preparation and characterization of biopolymer blends based on polyvinyl alcohol and starch. Roman Biotechnol Lett 20(20):10306–10315

39. Hejri Z, Seifkordi AA, Ahmadpour A, Zebarjad SM, Maskooki A (2013) Biodegradable starch/poly (vinyl alcohol) film reinforced with titanium dioxide nanoparticles. Int J Miner Metall Mater 20:1001–1011

40. Singha AS, Kapoor H (2014) Effects of plasticizer/cross-linker on the mechanical and thermal properties of starch/PVA blends. Iranian Polym J 23:655–662

41. Imam SH, Cinelli P, Gordon SH, Chiellini E (2005) Characterization of biodegradable composite films prepared from blends of poly(vinyl alcohol), cornstarch, and lignocellulosic fiber. J Polym Environ 13:47–55

42. Kopčilová M, Hubáčková J, Růžička J, Dvořáčková M, Julinová M, Koutný M, Tomalová M, Alexy P, Bugaj P, Filip J (2012) Biodegradability and mechanical properties of poly(vinyl alcohol)-based blend plastics prepared through extrusion method. J Polym Environ 21:88–94

43. Kale G, Kijchavengkul T, Auras R, Rubino M, Selke SE, Sher Paul Singh SP (2007) Compostability of bioplastic packaging materials: an overview. Macromol Biosci 7(3): 255–277

44. Jayasekara R, Harding I, Bowater I, Christie GBY, Lonergan GT (2004) Preparation, surface modification and characterisation of solution cast starch PVA blended films. Polym Testing 23(1):17–27

45. Mangaraj S, Yadav A, Bal LM, Dash SK, Mahanti NK (2018) Application of biodegradable polymers in food packaging industry: a comprehensive review. J Packag Technol Res 3(1): 77–96

46. Heidarian P, Behzad T, Sadeghi M (2017) Investigation of cross-linked PVA/starch biocomposites reinforced by cellulose nanofibrils isolated from aspen wood sawdust. Cellulose 24:3323–3339

47. Cano A, Fortunati E, Chafer M, Gonzalez-Martınez C, Chiralt A, Kenny JM (2015) Effect of cellulose nanocrystals on the properties of pea starch–poly(vinyl alcohol) blend films. J Mater Sci 50(21):6979–6992

48. Cano A, Fortunati E, Chafer M, Kenny JM, Chiralt A, Gonzalez-Martínez C (2015) Properties and ageing behaviour of pea starch films as affected by blend with poly(vinyl alcohol). Food Hydrocolloids 48:84–93

49. Qiu K, Netravali AN (2015) Polyvinyl alcohol based biodegradable polymer nanocomposites. In: Chu CC (ed) Biodegradable polymers. Nova Science, New York, pp 325–379

50. Sang BI, Hori K, Tanji Y, Unno H (2002) Fungal contribution to in situ biodegradation of poly(3-hydroxybutyrate-co-3-hydroxyvalerate) film in soil. Appl Microbiol Biotechnol 58:241–247

51. Talja RA, Helén H, Roos YH, Jouppila K (2007) Effect of various polyols and polyol contents on physical and mechanical properties of potato starch-based films. Carbohydr Polym 67:288–295

Chapter 5
Barrier Properties of PVA/HNT Bionanocomposite Films

Abstract Barrier properties in terms of water vapour transmission rate (WVTR), water vapour permeability (WVP), oxygen and air permeabilities have been evaluated for polyvinyl alcohol (PVA)/starch (ST)/glycerol (GL)/halloysite nanotube (HNT) bionanocomposites at different HNT contents of 0.25, 0.5, 1, 3 and 5 wt%, as compared with those of neat PVA and PVA blends. WVTR and WVP of bionanocomposites at ambient conditions decrease linearly by 52.34 and 73.59% with increasing the HNT content from 0 to 5 wt% due to their tortuosity effect when compared with those of neat PVA and PVA blends. Although WVTR and WVP of neat PVA and PVA blends increase remarkably with increasing the temperature from 25 to 55 °C and relative humidity (RH) level from 30 to 90%, WVTR and WVP of bionanocomposites show less sensitivity to the variation of temperature and RH level, resulting from the incorporation of HNTs with high thermal stability and moderate hydrophobicity. Furthermore, oxygen and air permeabilities of bionanocomposites decline significantly by 74.84 and 75.98%, as opposed to those of PVA blends at ambient conditions. Experimental permeability data show good agreement with Nielsen model for well-aligned HNT distribution when accurate values of HNT aspect ratios are employed based on experimental measurements via atomic force microscopy (AFM).

Keywords Bionanocomposites · Barrier properties · Permeability · Relative humidity (RH) · Theoretical models

The shelf life of foodstuffs is associated with barrier properties of packaging materials in order to protect foodstuffs from the dehydration and oxidation [1, 2]. Consequently, barrier properties should be considered for biopolymers and their nanocomposites when considered as food packaging materials. Moreover, such properties of biopolymers and their nanocomposites can be affected by changing the temperature and relative humidity (RH) [3]. Thus, the effects of temperature and RH gradient on the WVP was studied based on several material systems such as poly (hydroxy-butyrate) (PHB)/organo-modified montmorillonite (MMT) nanocomposite films, poly(hydroxyl-butyrate-co-hydroxy-valerate) (PHBHV)/organo-modified

© Springer Nature Singapore Pte Ltd. 2020
Z. W. Abdullah and Y. Dong, *Polyvinyl Alcohol/Halloysite Nanotube Bionanocomposites as Biodegradable Packaging Materials*,
https://doi.org/10.1007/978-981-15-7356-9_5

MMT nanocomposite films [4], PVA/MMT nanocomposite films [5], edible high amylose corn ST films [3], chitosan films [6] and protein films [7]. These investigations have demonstrated that material permeability increases linearly with increasing the temperature and RH gradient. In this chapter, barrier properties of neat PVA, PVA blends and bionanocomposite films were evaluated at ambient conditions, while the effects of temperature range between 25 and 55 °C and RH gradient between 10 to 70% on WVTR and WVP were thoroughly investigated. Furthermore, these results have been compared with theoretical models of permeability based on typical aspect ratios of as-received HNTs, as well as accurately calculated aspect ratios of embedded HNTs in bionanocomposite films via AFM.

5.1 Water Vapour Transmission and Water Vapour Permeability

The WVTR and WVP of neat PVA, PVA blends and PVA/ST/GL/HNT bionanocomposite films at different HNT contents were determined at 25 °C with a RH level of 50% ± 2%, as demonstrated in Fig. 5.1. The average data related to WVTR and WVP of neat PVA and PVA blends show relatively high values because of typical hydrophilic nature of such polymers with many hydroxyl groups, as reported elsewhere [8, 9]. The presence of GL and ST is the main reason behind the high WVTR and WVP of PVA blends compared with those of neat PVA films, which is associated with the improvements for the mobility of polymeric chains and hydrophilicity, respectively [9, 10]. The WVTR and WVP of PVA/GL blends increased by 7.08 and 16.72%, respectively, as opposed to those of neat PVA, in good accordance with other studies [11–14]. GL as a typical plasticiser enables to diminish the intermolecular interactions and promote the mobility of polymeric chains leading to the improvement of both WVTR and WVP [15]. Moreover, Talja et al. [16] stated that the presence of plasticiser could improve the diffusion rate of water molecules within polymers resulting in higher WVTR and WVP. Additionally, PVA/ST blends yield remarkable increases in WVTR and WVP by 21.27 and 30.96%, respectively, relative to those of neat PVA. Generally, the water solubility of PVA and water sensitivity of ST would produce water-sensitive and permeable blends [9]. The WVTR of PVA/ST/GL blends becomes lower than those of PVA/GL and PVA/ST blends, while the WVP of PVA/ST/GL blends has been identified between those of PVA/GL and PVA/ST blends due to the presence of GL to enhance the compatibility between PVA and ST, and can consume some hydrogel groups to build up hydrogen bonds.

The WVTR and WVP of PVA/ST/GL/HNT bionanocomposite films are reduced when compared with those of corresponding blend films since the addition of HNTs generates tortuous paths within bionanocomposite films [17] to promote better water resistance [18]. As such, WVTR and WVP of bionanocomposite films are

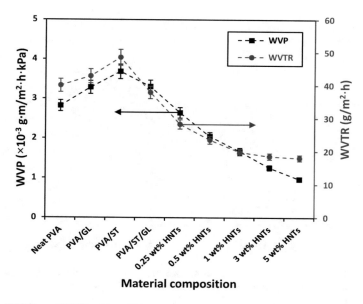

Fig. 5.1 WVTR and WVP of neat PVA, PVA blend and PVA/ST/GL/HNT bionanocomposite films at 25 °C and a RH level of 50%

decreased significantly by 52.34 and 73.59%, respectively, with increasing the HNT contents from 0 to 5 wt%. Lee et al. [19] reported a similar reduction in WVP of chitosan/HNT nanocomposites and chitosan/clove essential oil (CEO)/HNT nanocomposites by 16.11 and 15.67%, respectively, as compared with their blends due to the presence of HNTs to consume free hydroxyl groups in nanocomposites and create hydrogen bonding with polymer matrices. Such decreases in WVTR and WVP become less pronounced between 1 and 5 wt% HNTs, resulting from typical HNT agglomeration, which is in good agreement with previous findings on plasticised PVA/ST nanocomposites [20–24]. In polymer/clay nanocomposites, the agglomeration of nanofillers could generate newly connected pathways for permeable molecules at the polymer matrix/clay interfaces instead of the touristy resulting in the permeability improvement [25] shown in Fig. 5.2.

It is clearly seen from Fig. 5.3 that HNTs have significant effect on WVP of PVA/ST blends as compared with other types of nanofillers due to their moderate hydrophobicity, resulting from their lower number of hydroxyl groups on surfaces with reasonable water resistance. This is evidenced by the results of water contact angles mentioned in previous chapter. Overall, the improvement rate in barrier properties of nanocomposites can be determined by the dispersion of nanofillers to create tortuous paths for inhibiting gas/liquid permeation [26, 27].

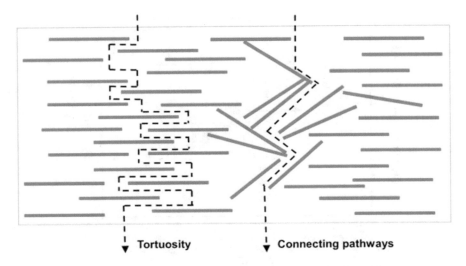

Fig. 5.2 Schematic diagram of permeation through the interfaces of polymer/clay nanocomposites. Image taken from [25] with copyright permission from Elsevier

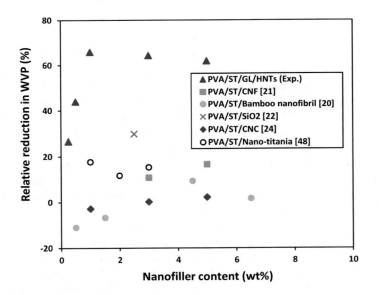

Fig. 5.3 Effect of different nanofiller types on WVP of PVA/ST nanocomposite films

5.1.1   Effect of Temperature

The food packaging materials may be used at different temperatures and RH levels during the transport and display processes. As such, WVTR and WVP of neat PVA, PVA blends and PVA/ST/GL/HNT bionanocomposite films were evaluated at 25, 35, 45 and 55 °C, as illustrated in Fig. 5.4a, b, respectively. There exists a linearly increasing trend for WVTR and WVP of all material films when the temperature is increased from 25 to 55 °C. It is clearly known that WVP values of polymers and corresponding nanocomposites are increased with increasing the temperature owing to enhancing free volume and the mobility of polymeric chains, as well as increasing the diffusion rate and energy level of permeable molecules [3, 28, 29]. However, the WVTR and WVP of neat PVA films are increased gradually with increasing the temperatures from 25 to 55 °C by 15.06 and 4.98%, respectively. The WVTR and WVP of neat PVA appear to be less influenced by the temperature variation when compared with those of PVA blends because the selected temperature range of 25–55 °C is still less than glass transition temperature (T_g) of PVA at 70.70 °C, as reported in Chap. 3. For the same reason, WVTR and WVP of PVA/GL blends, PVA/ST/GL blends and PVA/ST/GL/HNT bionanocomposites were moderately increased with increasing the temperature from 25 to 35 °C, while such an increasing trend becomes more pronounced beyond 35°C. For instance, the WVTR and WVP of PVA/GL blends are increased slightly by 10.56 and 6.70%, respectively at 35 °C, which is followed by a remarkable increase with increasing the temperature from 35 to 55 °C. The significant increases in WVTR and WVP of PVA/GL blends are related to the selected temperature range to be much closer to their T_g of 47.70 °C. This is also the case for PVA/ST/GL blends and their corresponding bionanocomposite films. Whereas, WVTR and WVP of PVA/ST blends instead demonstrate a gradually increasing tendency over the selected temperature range despite being still less than the T_g of 74.10 °C. At all temperature levels, PVA/ST/GL/HNT bionanocomposite films have lower WVTR and WVP than those of corresponding blends due to the presence of HNTs to undermine the temperature-dependent effect. In other words, the increasing rates in WVTR and WVP of bionanocomposite films in terms of temperature level are lower than those of blends. In a similar manner, Huang et al. [5] reported that WVP of PVA/MMT bionanocomposite films was less dependent on the temperature level because the incorporation of nanofillers could restrict chain mobility of polymeric molecules, and further reduce the diffusion rate of water molecules as the indicator for lower WVTR and WVP.

This temperature dependence on WVP of PVA/ST/GL blends, and their corresponding bionanocomposite films can be described according to Arrhenius equation as follows [14, 28, 31–33]:

$$\text{WVP} = P_a \exp\left(\frac{-E_p}{RT}\right) \tag{5.1}$$

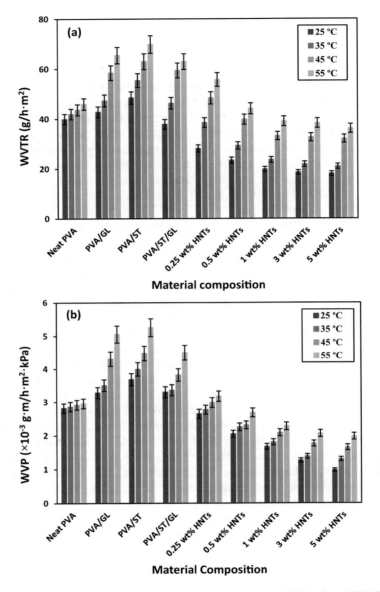

Fig. 5.4 Temperature effect on **a** WVTR and **b** WVP of neat PVA, PVA blends and PVA/ST/GL/HNT bionanocomposite films at different HNT contents. Figures taken from [30] with copyright permission from Elsevier

where P_a is the Arrhenius constant (i.e. pre-exponential constant), E_p is the activation energy of permeability (kJ/mol), R is the universal gas constant (8.314 kJ/mol K) and T is the absolute temperature (K). The logarithmic relationship between

WVP values of PVA/ST/GL blends, as well as their corresponding bionanocomposite films and $1/T$ shows good line fitting with Arrhenius equation, as demonstrated in Fig. 5.5. The linear gradient of this relationship means the high WVP of material depends on temperature, which is evidently revealed for PVA/ST/GL blends with their increasing line gradient when increasing the temperature level. Whereas, the high thermal stability of HNTs [34] clearly reduces the effect of temperature dependence on bionanocomposite films at the HNT contents of 0.25–1 wt%, as compared with that of PVA/ST/GL blends alone, in sign of less gradient lines. The lines gradient is increased again for bionanocomposites reinforced with 3 and 5 wt% HNTs due to the HNT agglomeration issue to hinder the effect of HNTs on WVP.

According to Huang et al. [5], E_p is defined as the minimum energy required by permeable molecules to overcome interaction forces between material molecules and diffuse through the materials. Consequently, the high value of E_p indicates that permeable molecules need higher energy to diffuse through the materials. In other words, a material possesses a low permeability rate at the high E_p value leading to less sensitivity to different temperature levels [7, 35]. It is clearly shown in Table 5.1 that PVA blends have lower E_p than that of neat PVA. Accordingly, WVP of neat PVA is less sensitive to temperature change as opposed to that of PVA blends. Additionally, PVA/ST/GL/HNT bionanocomposites have higher E_p values compared with that of corresponding blends, which are increased in a

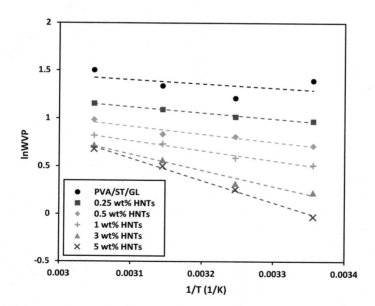

Fig. 5.5 Arrhenius relationship between WVP of PVA/ST/GL blends and corresponding bionanocomposite films at different HNT contents and temperature levels. Figure taken from [30] with copyright permission from Elsevier

Table 5.1 Activation energies of permeation and Arrhenius constants of neat PVA, PVA blends and PVA/ST/GL/HNT bionanocomposite films

Material composition	E_p (kJ/mol)	P_a
Neat PVA	11.27	2.97
PVA/GL	8.46	5.05
PVA/ST	7.97	5.47
PVA/ST/GL	10.84	4.52
PVA/ST/GL/0.25 wt% HNTs	12.01	3.26
PVA/ST/GL/0.5 wt% HNTs	12.98	2.71
PVA/ST/GL/1 wt% HNTs	13.04	2.43
PVA/ST/GL/3 wt% HNTs	14.13	2.48
PVA/ST/GL/5 wt% HNTs	14.16	2.22

Table taken from [30] with copyright permission from Elsevier

monotonic manner with increasing the HNT content, also evidenced in other studies [4, 5, 35]. Overall, the incorporation of HNTs within PVA/ST/GL blends in bionanocomposite films greatly reduces WVTR and WVP of bionanocomposite films in the temperature range of 25–55 °C according to Arrhenius relationship.

5.1.2   Effect of RH Gradient

WVTR and WVP of neat PVA, PVA blends and PVA/ST/GL/HNT bionanocomposite films were evaluated at different RH gradients of 10, 30, 50 and 70% ± 2% at the room temperature of 25 °C, as demonstrated in Fig. 5.6a, b, respectively. According to Fick's first law of diffusion, the flux (F_x) in one direction (∂_x) is proportional to the gradient of concentration (∂_c) in the same direction shown below [28, 36]:

$$F_x = -D\left(\frac{\partial_c}{\partial_x}\right) \tag{5.2}$$

Accordingly, the RH gradient is the driving force of permeation process across film thickness. It is evidently shown that WVTR and WVP of neat PVA, PVA blends and PVA/ST/GL/HNT bionanocomposites are increased remarkably with increasing RH gradient, which is in good agreement with previous studies [6, 16]. Consequently, the higher values of WVTR and WVP were recorded at the RH gradient of 70% ± 2% as opposed to those recorded at 10% ± 2%. According to Cuq et al. [37], the plasticisation effect of water molecules could appear at the high RH level leading to the improvement of free volume of polymeric molecules while the diffusion and transfer of water molecules can be improved across the films. Moreover, Ashley [31] and Mo et al. [38] stated that hydrophilic polymers like PVA had increased WVTR and WVP with the RH level due to the interaction of water molecules with hydroxyl groups of polymers through hydrogen bonding

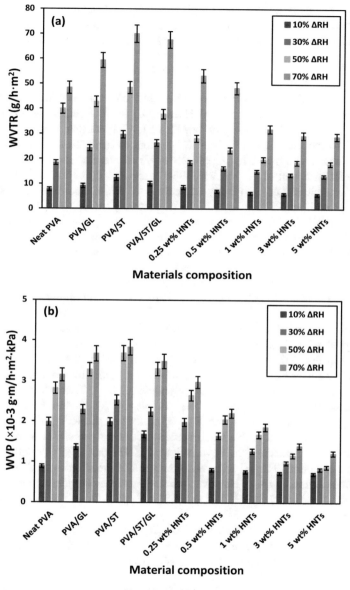

Fig. 5.6 RH gradient effect on **a** WVTR and **b** WVP of neat PVA, PVA blend and PVA/ST/GL/ HNT bionanocomposite films at different HNT contents. Figures taken from [30] with copyright permission from Elsevier

resulting in the increase in water solubility coefficient (S). On the other hand, water diffusivity coefficient (D) may also be increased owing to the plasticisation effect of water molecules to enhance the permeability coefficient (P) as follows [7, 31, 38]:

$$P = S \times D \qquad\qquad (5.3)$$

As such, WVTR and WVP of neat PVA are increased by 518.79 and 257.95%, respectively, with increasing the ΔRH from 10 to 70% ± 2%. At different RH gradients, PVA blends have higher WVTR and WVP compared with those of neat PVA due to the presence of GL and ST leading to better chain mobility of polymeric molecules and higher hydrophilicity of blends, as discussed earlier [9, 10]. At the same RH gradient of 10% ± 2%, WVTR and WVP of PVA/GL, PVA/ST and PVA/ST/GL blends are evidently increased by 17.90 and 54.55%, 60.01 and 125.0%, as well as 30.05 and 89.77%, respectively, as opposed to those of neat PVA. A similar trend was recorded at other RH gradients (i.e. 30, 50 and 70% ± 2%). These findings are in good accordance with other ST nanocomposites [16]. Overall, WVTR and WVP of neat PVA are increased linearly with increasing RH gradient due to the plasticisation effect of water molecules. Additionally, this increasing trend becomes more pronounced for PVA blends due to the presence of GL and ST for better mobility of polymeric molecules and higher hydrophilicity of blends.

WVTR and WVP of PVA/ST/GL/HNT bionanocomposite films are greatly decreased at all RH gradients compared with those of corresponding blends, as illustrated in Fig. 5.6a, b, respectively. The incorporation of HNTs within polymer matrices generates tortuous paths leading to the restriction to both diffusion of permeable molecules and mobility of polymeric chains [17, 26]. WVTR and WVP of bionanocomposite films decline remarkably by 57.28 and 65.26% with the increasing HNT content from 0 to 5 wt%, as well as the RH gradient from 10 to 70% ± 2%. Additionally, this improvement in barrier properties of bionanocomposite films diminishes slightly beyond 1 wt% HNTs due to HNT agglomeration at the HNT contents of 3 and 5 wt%, according to our predetermined SEM and AFM results presented in Chap. 3. However, barrier properties of bionanocomposite films reinforced with 3 and 5 wt% HNTs are still better than those of PVA/ST/GL blends at all RH gradients.

Depending on WVP data of PVA/ST/GL blends and corresponding bionanocomposite films, a clear linear relationship could be established between WVP and RH gradient (ΔRH) as follows (see Fig. 5.7):

$$WVP = a\Delta RH + b \qquad\qquad (5.4)$$

where a and b are constants for line fitting. Furthermore, the slopes of these lines are decreased with increasing the HNT content. In other words, the WVP of bionanocomposite films becomes less sensitive to increasing the RH gradient as

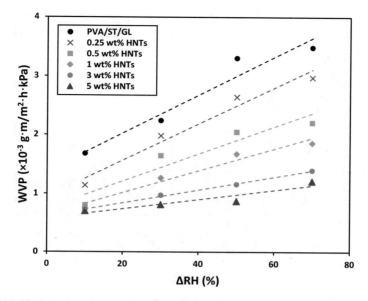

Fig. 5.7 Relationship between WVP and RH gradient for PVA/ST/GL blend and corresponding bionanocomposite films at different HNT contents. Figure taken from [30] with copyright permission from Elsevier

opposed to that of PVA/ST/GL blends. This phenomenon is ascribed to the improvement of relative surface hydrophobicity in the presence of HNTs in good accordance with the results of water contact angles obtained in Chap. 4.

5.2   Gas Permeability

Gas permeability of neat PVA, PVA blend and PVA/ST/GL/HNT bionanocomposite films was investigated by using oxygen and air, Fig. 5.8. The good barrier properties mean lower permeability of materials (harder to be permeated by permeable molecules) [39]. It can be clearly seen that the gas permeability of all material films appears to be lower than WVP because polar polymers like PVA have good gas barrier properties, but poor water resistance due to the presence of many hydroxyl groups, while their gas barrier properties are decreased in the presence of plasticisers [31]. Similarly, Lim et al. [8] has found that high crystallinity and strong intermolecular interactions of PVA give rise to good oxygen barrier properties though lower water resistance arises from the presence of many hydroxyl groups. A similar trend has been reported for oxygen and air permeabilities of all material films because there is no direct interaction between polymeric molecules and gas molecules, as evidenced in previous work [40, 41]. The addition of GL increases oxygen and air permeabilities of PVA/GL blends by 77.59

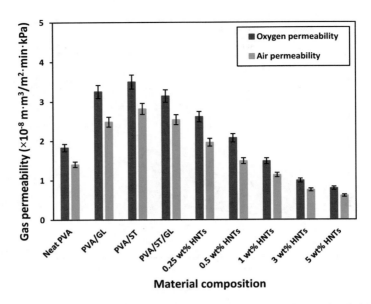

Fig. 5.8 Oxygen and air permeabilities of neat PVA, PVA blend and PVA/ST/GL/HNT bionanocomposite films at different HNT contents. Figure taken from [30] with copyright permission from Elsevier

and 75.88%, respectively, as opposed to those of neat PVA due to the plasticisation effect of GL. The presence of plasticisers improves the mobility of polymeric chains and increases free volume of polymers that can be easily penetrated by permeable molecules with poor barrier properties. A further increase for oxygen and air permeabilities of PVA/ST blends by 91.25 and 99.29%, respectively, is also shown when compared with those of neat PVA owing to partial compatibility between PVA and ST in the absence of plasticiser. This results from the diffusion improvement of permeable molecules due to inhomogeneous morphological structures with several diffusion paths [42]. Consequently, oxygen and air permeabilities of PVA/ST/GL blends are reduced by 10.28 and 9.62%, respectively, relative to PVA/ST blends due to the addition of GL.

In general, when impermeable fillers are incorporated within polymer matrices, permeable molecules should travel around these fillers in much longer tortuous paths to be permeable through composites particularly when there is no interaction taking place between permeable molecules (e.g. oxygen and air) and matrix constituents [43]. Such tortuous paths can be described by tortuosity factor (τ) depending on simple Nielsen model given below [18, 41, 44, 45]:

$$\frac{P_c}{P_o} = \frac{1 - \emptyset}{\tau} \qquad (5.5)$$

Table 5.2 Relative oxygen permeabilities and tortuosity of bionanocomposite films at different HNT contents

HNT content (wt%)	P_c/P_o	τ
0	1.00	1.00
0.25	0.83	1.18
0.5	0.66	1.47
1	0.47	2.00
3	0.32	2.59
5	0.25	2.60

Table taken from [30] with copyright permission from Elsevier

where P_c, P_o and $\emptyset$ are the permeabilities of nanocomposites and polymer matrices, respectively, as well as the volume fraction of impermeable nanofillers [41, 45]. It is clearly revealed that τ values are increased in a monotonic manner with increasing the HNT content from 0 to 5 wt%, as listed in Table 5.2. This trend could be interpreted by the generation of tortuous paths within PVA/ST/GL/HNT bionanocomposite films leading to much better barrier properties.

As such, oxygen and air permeabilities of bionanocomposite films are decreased significantly by 74.84 and 75.98%, respectively, with increasing the HNT content from 0 to 5 wt% in good accordance with other studies [18, 44, 46, 47]. This reduction trend in oxygen and air permeabilities becomes less pronounced beyond 1 wt% HNTs owing to their agglomeration to generate direct paths for permeable molecules rather than tortuous paths mentioned previously [21, 48]. Such results coincide with other findings based on PVA/ST/nanotitania nanocomposites [49]. It is manifested that morphological structures of nanocomposites in terms of nanofiller dispersion play an important role in the improvement of barrier properties [50].

5.3   Comparison Between Experimental Data and Theoretical Models

The most popular Nielsen model and Cussler model have been considered to predict water vapour and gas permeabilities of polymer nanocomposites reinforced with ribbon-like nanofillers [25, 51]. There is no definite model for tubular nanofillers, and ribbon-like structures are the closest to HNTs in shape. Consequently, these two models based on ribbon-like nanofillers in terms of regular and random dispersion of nanofillers have been used to predict water vapour, oxygen and air permeabilities of PVA/ST/GL/HNT bionanocomposite films in comparison with their experimental data.

Nielsen model for regular and random nanofiller dispersion can be written in Eqs. (5.6) and (5.7), respectively [25, 26, 51, 52]:

$$\frac{P_c}{P_o} = \frac{1 - \emptyset}{1 + \left(\frac{\alpha}{2}\right)\emptyset} \tag{5.6}$$

$$\frac{P_c}{P_o} = \frac{1 - \emptyset}{1 + \frac{1}{3}\left(\frac{\alpha}{2}\right)\emptyset} \tag{5.7}$$

where α is aspect ratio of HNTs (i.e. $\alpha = L/D$). $\emptyset$ can be calculated from the following equation [41, 53]:

$$\frac{1}{\emptyset} = 1 + \frac{\rho_i(1 - \mu_i)}{\rho_p \mu_i} \tag{5.8}$$

where ρ_i and μ_i are the density (i.e. $\rho_i = 2.53$ g/cm^3 for HNTs based on HNT data sheet) and weight fraction of impermeable phase (i.e. HNTs in this case), respectively. Whereas, ρ_p is the density of permeable phase (i.e. polymer blend matrices) that can be calculated as follows [54]:

$$\rho_p = \sum_{x=1}^{n} \rho_x V_x \tag{5.9}$$

where ρ_x and V_x are the density and volume fraction of each component in polymer blend matrices (i.e. PVA, ST and GL) [54]. Typical aspect ratio of as-received HNTs ($\alpha = 39.22$) and variable aspect ratios of embedded HNTs in PVA/ST/GL/HNT bionanocomposite films were determined using AFM (see Chap. 3) so that they could be further employed in Nielsen model and Cussler model for the best fitting with experimental data, as illustrated in Fig. 5.9a, b, respectively. When the typical aspect ratio of as-received HNTs is considered, good agreement has been detected between Nielsen model for regular nanofiller dispersion and relative permeability experimental data of bionanocomposites up to 1 wt% HNTs. This behaviour could be interpreted in terms of significant reduction in relative permeabilities of bionanocomposites due to good dispersion of HNTs in the range of 0–1 wt%. Conversely, Nielsen model for random nanofiller dispersion can fit better beyond 1 wt% HNTs accordingly, which is believed to be associated with the HNT agglomeration according to our SEM and AFM images obtained in Chap. 3. In general, most developed permeability models for nanocomposite films are based on regular shapes and sizes of nanofillers within polymer matrices [26]. In fact, the shapes and sizes of nanofillers can be greatly altered during the manufacturing steps of nanocomposites. Additionally, nanofillers may be clustered or agglomerated particularly at the high contents due to their weak van der Waals interactions. As such, it is worthwhile to use experimentally determined aspect ratios of embedded HNTs in relation to their different nanofiller contents in bionanocomposites for more accurate modelling work. When variable aspect ratios of HNTs are taken into account, the results based on Nielsen model for regular nanofiller dispersion have been found to show better agreement with experimental permeability data with the

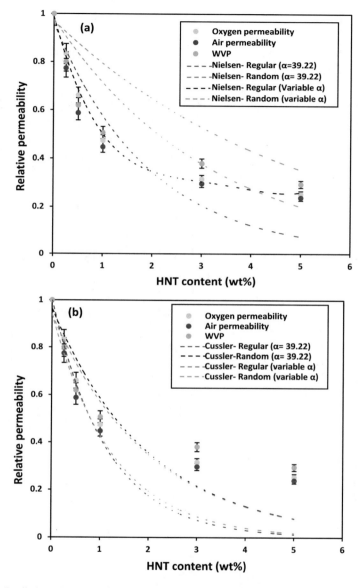

Fig. 5.9 Prediction of relative permeabilities of bionanocomposites using **a** Nielsen model and **b** Cussler model. Figures taken from [30] with copyright permission from Elsevier

consideration of HNT agglomeration, as depicted in Fig. 5.9a with lower error percentages listed in Table 5.3.

On the other hand, Cussler model for regular and random nanofiller dispersion can be written below according to Eqs. (5.10) and (5.11), respectively [26, 51, 52]:

Table 5.3 Data comparison between experimental and theoretical permeabilities based on Nielsen models for regular and random dispersion

HNTs (wt%)	Experimental	Regular, α = 39.22	Error (%)	Regular, variable α	Error (%)	Random, α = 39.22	Error (%)	Random, variable α	Error (%)
WVP									
0	1	1	0	1	0	1	0	1	0
0.25	0.79	0.78	2.01	0.78	2.14	0.90	11.91	0.90	11.91
0.5	0.61	0.63	3.08	0.71	13.38	0.82	25.29	0.86	28.73
1	0.50	0.45	11.29	0.49	2.71	0.69	27.24	0.78	35.49
3	0.38	0.17	113.48	0.30	24.59	0.37	2.15	0.52	27.48
5	0.29	0.08	245.88	0.26	13.07	0.20	44.82	0.35	17.18
Air permeability									
0	1	1	0	1	0	1	0	1	0
0.25	0.77	0.78	1.31	0.78	1.18	0.90	14.79	0.90	14.79
0.5	0.58	0.63	8.05	0.71	17.83	0.82	29.13	0.86	32.39
1	0.44	0.45	1.73	0.49	9.31	0.69	35.76	0.78	43.04
3	0.29	0.17	66.40	0.30	2.88	0.37	20.37	0.52	43.47
5	0.23	0.08	180.47	0.26	8.30	0.20	17.43	0.35	32.84
Oxygen permeability									
0	1	1	0	1	0	1	0	1	0
0.25	0.83	0.78	6.22	0.78	6.36	0.90	8.28	0.90	8.28
0.5	0.66	0.63	3.39	0.71	7.59	0.82	20.30	0.86	23.97
1	0.47	0.45	4.10	0.49	3.91	0.69	31.93	0.78	39.65
3	0.31	0.17	77.35	0.30	3.50	0.37	15.13	0.52	39.75
5	0.25	0.08	197.78	0.26	2.64	0.20	24.68	0.35	28.69

$$\frac{P_c}{P_o} = \frac{1 - \emptyset}{1 + \left(\frac{\alpha\emptyset}{2}\right)^2} \tag{5.10}$$

$$\frac{P_c}{P_o} = \frac{1 - \emptyset}{\left(1 + \frac{\alpha\emptyset}{3}\right)^2} \tag{5.11}$$

It is clearly shown in Fig. 5.9b and Table 5.4 that Cussler model for regular and random dispersion of nanofillers yields good fitting with permeability experimental data of bionanocomposites based on the aspect ratios of both as-received HNTs and embedded HNTs within polymer matrices in bionanocomposites at the HNT contents of 0–1 wt%. Nonetheless, beyond 1 wt% HNTs, Cussler model fails to fit experimental data of bionanocomposites owing to its applicability to nanocomposite systems at low nanofiller contents with high aspect ratios of nanofillers [18, 47]. Overall, nanofiller content, nanofiller aspect ratio and nanofiller dispersion within polymer matrices in nanocomposite systems should be carefully considered for the effective prediction in the permeabilities of bionanocomposite films.

5.4   Summary

Barrier properties in terms of water vapour and gas permeabilities play an important role in the selection of nanocomposite systems for food packaging applications. Neat PVA as a typical water-soluble polymer possesses good gas barrier properties but poor water resistance. Furthermore, the addition of GL slightly reduces WVTR and WVP owing to the consumption of some free hydroxyl groups through hydrogen bonding. Clearly, the increases in both WVTR and WVP were reported for PVA/ST blends due to the hydrophilic characteristic of ST. It is worth mentioning that WVTR and WVP of PVA/ST/GL blends are ranging between those of PVA/GL blends and PVA/ST blends. On the other hand, a significant reduction in WVTR and WVP was observed for PVA/ST/GL/HNT bionanocomposites, which is ascribed to the inclusion of HNTs with moderate hydrophobicity to restrict the diffusion of water molecules. Increasing the temperature from 25 to 55 °C at a RH level of 50% causes the increases in WVTR and WVP of neat PVA and PVA blends since the chain mobility of polymeric molecules tend to improve resulting in better diffusion of water molecules. On the other hand, bionanocomposite films become less sensitive to increasing the temperature according to Arrhenius relationship associated with the reduction of mass and heat transfer in the presence of HNTs. A similar increasing trend is clearly demonstrated as well in WVTR and WVP with increasing ΔRH from 10 to 70%. This is because ΔRH is deemed as the driving force in the permeation process despite being less influential to bionanocomposite films as compared to PVA blends. A similar trend was reported for air and oxygen permeabilities with the clear reduction in gas permeability in the

Table 5.4 Data comparison between experimental and theoretical permeabilities based on Cussler models for regular and random dispersion

HNTs (wt %)	Experimental	Regular, α = 39.22	Error (%)	Regular, variable α	Error (%)	Random, α = 39.22	Error (%)	Random, variable α	Error (%)
WVP									
0	1	1	0	1	0	1	0	1	0
0.25	0.79	0.92	13.44	0.82	2.93	0.71	11.39	0.71	11.54
0.5	0.61	0.76	18.83	0.59	4.78	0.53	15.75	0.63	2.00
1	0.50	0.43	15.61	0.42	19.15	0.32	57.75	0.47	6.61
3	0.38	0.05	555.17	0.21	77.57	0.07	435.21	0.18	108.79
5	0.29	0.01	2000.0	0.08	237.93	0.02	1236.36	0.09	226.66
Air permeability									
0	1	1	0	1	0	1	0	1	0
0.25	0.77	0.92	16.26	0.82	6.10	0.71	7.75	0.71	7.90
0.5	0.58	0.76	23.00	0.59	0.59	0.53	9.81	0.63	7.04
1	0.44	0.43	2.07	0.42	5.2	0.32	39.28	0.47	5.87
3	0.29	0.05	410.68	0.21	38.41	0.07	317.18	0.18	62.74
5	0.23	0.01	1602.85	0.08	174.02	0.02	983.63	0.09	164.88
Oxygen permeability									
0	1	1	0	1	0	1	0	1	0
0.25	0.83	0.92	9.87	0.82	1.06	0.71	15.98	0.71	16.72
0.5	0.66	0.76	13.40	0.59	11.79	0.53	23.49	0.63	33.93
1	0.47	0.43	8.15	0.42	11.45	0.32	47.57	0.47	52.63
3	0.31	0.05	444.31	0.21	47.52	0.07	344.64	0.18	68.43
5	0.25	0.01	1707.97	0.08	190.93	0.02	1050.53	0.09	74.68

presence of HNTs on account of a remarkable increase in tortuosity factor when incorporated with HNTs in bionanocomposite films. The experimental data of bionanocomposite permeabilities have the best agreement with Nielsen model for regular dispersion of HNTs with variable aspect ratios, while Cussler model for regular and random dispersions of nanofillers can also fit the experimental data at HNT contents of 0–1 wt%.

References

1. Othman SH (2014) Bio-nanocomposite materials for food packaging applications: types of biopolymer and nano-sized filler. Agricul Agricul Sci Procedia 2:296–303
2. Mangaraj S, Yadav A, Bal LM, Dash SK, Mahanti NK (2018) Application of biodegradable polymers in food packaging industry: a comprehensive review. J Packag Technol Res 3 (1):77–96
3. Bertuzzi MA, Vidaurre EFC, Armada M, Gottifredi JC (2007) Water vapor permeability of edible starch based films. J Food Eng 80(3):972–978
4. Akin O, Tihminlioglu F (2017) Effects of organo-modified clay addition and temperature on the water vapor barrier properties of polyhydroxy butyrate homo and copolymer nanocomposite films for packaging applications. J Polym Environ 26(3):1121–1132
5. Huang JY, Limqueco J, Chieng YY, Li X, Zhou W (2017) Performance evaluation of a novel food packaging material based on clay/polyvinyl alcohol nanocomposite. Innovat Food Sci Emerg Technol 43:216–222
6. Wiles JL, Vergano PJ, Barron FH, Bunn JM, Testin RF (2000) Water vapor transmission rates and sorption behavior of chitosan films. J Food Sci 65(7):1175–1179
7. Gennadios A, Brandenburg AH, Park JW, Weller CL, Testin RF (1994) Water vapor permeability of wheat gluten and soy protein isolate films. Ind Crop Prod 2:189–195
8. Lim M, Kwon H, Kim D, Seo J, Han H, Khan SB (2015) Highly-enhanced water resistant and oxygen barrier properties of cross-linked poly(vinyl alcohol) hybrid films for packaging applications. Prog Organic Coating 85:68–75
9. Cano A, Fortunati E, Chafer M, Kenny JM, Chiralt A, Gonzalez-Martínez C (2015) Properties and ageing behaviour of pea starch films as affected by blend with poly(vinyl alcohol). Food Hydrocolloids 48:84–93
10. Cano AI, Cháfer M, Chiralt A, González-Martínez C (2015) Physical and microstructural properties of biodegradable films based on pea starch and PVA. J Food Eng 167:59–64
11. Jiang X, Luo Y, Hou L, Zhao Y (2016) The Effect of glycerol on the crystalline, thermal, and tensile properties of CaCl$_2$-doped starch/PVA films. Polym Compos 37(11):3191–3199
12. Imam SH, Cinelli P, Gordon SH, Chiellini E (2005) Characterization of biodegradable composite films prepared from blends of poly(vinyl alcohol), cornstarch, and lignocellulosic fiber. J Polym Environ 13:47–55
13. Zhang Y, Han JH (2006) Mechanical and thermal characteristics of pea starch films plasticized with monosaccharides and polyols. J Food Sci 71(2):109–118
14. Arvanitoyannis I, Nakayama A, Aiba S (1998) Edible films made from hydroxypropyl starch and gelatin and plasticized by polyols and water. Carbohydr Polym 36:105–119
15. Ismail H, Zaaba NF (2011) Effect of additives on properties of polyvinyl alcohol (PVA)/ tapioca starch biodegradable films. Polym Plast Technol Eng 50(12):1214–1219
16. Talja RA, Helén H, Roos YH, Jouppila K (2007) Effect of various polyols and polyol contents on physical and mechanical properties of potato starch-based films. Carbohydr Polym 67:288–295

17. Noshirvani N, Ghanbarzadeh B, Fasihi H, Almasi H (2016) Starch–PVA nanocomposite film incorporated with cellulose nanocrystals and MMT: a comparative study. Int J Food Eng 12 (1):37–48

18. Sadegh-Hassani F, Nafchi AM (2014) Preparation and characterization of bionanocomposite films based on potato starch/halloysite nanoclay. Int J Biol Macromol 67:458–462

19. Lee MH, Kim SY, Park HJ (2018) Effect of halloysite nanoclay on the physical, mechanical, and antioxidant properties of chitosan films incorporated with clove essential oil. Food Hydrocoll 84:58–67

20. Guimarães M Jr, Botaro VR, Novack KM, Teixeira FG, Tonoli GHD (2015) Starch/ PVA-based nanocomposites reinforced with bamboo nanofibrils. Ind Crop Prod 70:72–83

21. Heidarian P, Behzad T, Sadeghi M (2017) Investigation of cross-linked PVA/starch biocomposites reinforced by cellulose nanofibrils isolated from aspen wood sawdust. Cellulose 24:3323–3339

22. Tang S, Zou P, Xiong H, Tang H (2008) Effect of nano-SiO_2 on the performance of starch/ polyvinyl alcohol blend films. Carbohydr Polym 72(3):521–526

23. Tang X, Alavi S (2012) Structure and physical properties of starch/poly vinyl alcohol/laponite RD nanocomposite films. J Agricul Food Chemist 60(8):1954–1962

24. Cano A, Fortunati E, Chafer M, Gonzalez-Martinez C, Chiralt A, Kenny JM (2015) Effect of cellulose nanocrystals on the properties of pea starch–poly(vinyl alcohol) blend films. J Mater Sci 50(21):6979–6992

25. Tan B, Thomas NL (2016) A review of the water barrier properties of polymer/clay and polymer/graphene nanocomposites. J Memb Sci 514:595–612

26. Choudalakis G, Gotsis AD (2009) Permeability of polymer/clay nanocomposites: a review. Eur Polym J 45(4):967–984

27. Ghanbarzadeh B, Almasi H, Entezami AA (2011) Improving the barrier and mechanical properties of corn starch-based edible films: effect of citric acid and carboxymethyl cellulose. Ind Crop Prod 33(1):229–235

28. Rogers CE (1985) Permeation of Gases and Vapours in Polymers. In: Comyn J (ed) Polymer permeability. Elsevier Applied Science Publishers, London, pp 11–73

29. Poley LH, Silva MG, Vargas H, Siqueira MO, Sánchez R (2005) Water and vapor permeability at different temperatures of poly (3-hydroxybutyrate) dense membranes. Polímeros 15(1):22–26

30. Abdullah ZW, Dong Y, Han N, Liu S (2019) Water and gas barrier properties of polyvinyl alcohol (PVA)/starch (ST)/glycerol (GL)/halloysite nanotube (HNT) bionanocomposite films: experimental characterisation and modelling approach. Compos B Eng 174:107033

31. Ashley RJ (1985) Permeability and plastics packaping. In: Comyn J (ed) Polymer permeability. Elsevier Applied Science Publishers, London, pp 269–308

32. Siracusa V (2012) Food packaging permeability behaviour: a report. Int J Polym Sci 2012:302029

33. Morillon V, Debeaufort F, Blond G, Voilley A (2000) Temperature influence on moisture transfer through synthetic film. J Memb Sci 168:223–231

34. Yuan P, Tan D, Annabi-Bergaya F (2015) Properties and applications of halloysite nanotubes: recent research advances and future prospects. Appl Clay Sci 112–113:75–93

35. Gennadios A, Weller CL, Testin RF (1993) Temperature effect on oxygen permeability of edible protein-based films. J Food Sci 58(1):212–214

36. Comyn J (1985) Introduction to polymer permeability and the mathematics of diffusion. In: Comyn J (ed) Polymer permeability. Elsevier Applied Science Publishers, London, pp 1–10

37. Cuq B, Gontard N, Aymard C, Guilbert S (1997) Relative humidity and temperature effects on mechanical and water vapor barrier properties of myofibrillar protein-based films. Polym Gel Network 5(1):1–15

38. Mo C, Yuan W, Lei W, Shijiu Y (2014) Effects of temperature and humidity on the barrier properties of biaxially-oriented polypropylene and polyvinyl alcohol films. J Appl Packag Res 6(1):40–46

39. Feldman D (2013) Polymer nanocomposite barriers. J Macromol Sci A 50(4):441–448

40. Sridhar V, Tripathy DK (2006) Barrier properties of chlorobutyl nanoclay composites. J Appl Polym Sci 101(6):3630–3637

41. Picard E, Vermogen A, Gérard JF, Espuche E (2007) Barrier properties of nylon 6-montmorillonite nanocomposite membranes prepared by melt blending: influence of the clay content and dispersion state consequences on modelling. J Membrane Sci 292(1–2): 133–144

42. Zou GX, Ping-Qu J, Liang-Zou X (2008) Extruded starch/PVA composites: water resistance, thermal properties, and morphology. J Elastom Plast 40(4):303–316

43. Liu H, Bandyopadhyay P, Kim NH, Moon B, Lee JH (2016) Surface modified graphene oxide/poly(vinyl alcohol) composite for enhanced hydrogen gas barrier film. Polym Test 50:49–56

44. Alipoormazandarani N, Ghazihoseini S, Nafchi AM (2015) Preparation and characterization of novel bionanocomposite based on soluble soybean polysaccharide and halloysite nanoclay. Carbohyd Polym 134:745–751

45. Nielsen LE (1967) Models for the permeability of filled polymer systems. J Macromol Sci A Chemist 1(5):929–942

46. Liu D, Bian Q, Li Y, Wang Y, Xiang A, Tian H (2016) Effect of oxidation degrees of graphene oxide on the structure and properties of poly (vinyl alcohol) composite films. Compos Sci Technol 129:146–152

47. Huang HD, Ren PG, Chen J, Zhang WQ, Ji X, Li ZM (2012) High barrier graphene oxide nanosheet/poly(vinyl alcohol) nanocomposite films. J Memb Sci 409–410:156–163

48. Nafchi AM, Nassiri R, Sheibani S, Ariffin F, Karim AA (2013) Preparation and characterization of bionanocomposite films filled with nanorod-rich zinc oxide. Carbohydr Polym 96(1):233–239

49. Li D, Huang Y, Liu Y, Luo T, Xing B, Yang Y, Yang Z, Wu Z, Chen H, Zhang Q, Qin W (2018) Physico-mechanical and structural characteristics of starch/polyvinyl alcohol/ nano-titania photocatalytic antimicrobial composite films. LWT Food Sci Technol 96: 704–712

50. Cui Y, Kumar S, Konac BR, Houcke D (2015) Gas barrier properties of polymer/clay nanocomposites. RSC Adv 5(78):63669–63690

51. Takahashi S, Goldberg HA, Feeney CA, Karim DP, Farrell M, O'Leary K, Paul DR (2006) Gas barrier properties of butyl rubber/vermiculite nanocomposite coatings. Polymer 47 (9):3083–3093

52. Saritha A, Joseph K, Thomas S, Muraleekrishnan R (2012) Chlorobutyl rubber nanocomposites as effective gas and VOC barrier materials. Compos A Appl Sci Manuf 43(6):864–870

53. Alexandre B, Langevina D, Médéric P, Aubry T, Couderc H, Nguyen QT, Saiter A, Marais S (2009) Water barrier properties of polyamide 12/montmorillonite nanocomposite membranes: structure and volume fraction effects. J Memb Sci 328(1–2):186–204

54. Callister WD (2007) Materials science and engineering: an introduction (7th ed). Wiley, USA

Chapter 6
Component Migration of PVA/HNT Bionanocomposite Films

Abstract Polyvinyl alcohol (PVA)/starch (ST)/glycerol (GL)/halloysite nanotube (HNT) bionanocomposites at HNT contents of 0.25, 0.5, 1, 3 and 5 wt% were tested as food packaging materials based on their migration rates after contacting food simulants. The overall migration rates of PVA/ST/GL blends and their bio-nanocomposites in hydrophilic simulant exceed the standard overall migration limit due to their solubility in hydrophilic media, while the migration rates in acidic and lipidic food simulants are still much lower than the standard overall migration limit (OML) due to their limited interactions with similar media. Finally, PVA/ST/GL/ HNT bionanocomposites have been identified to be successfully used to pack peaches and freshly cut avocados with lower weight losses by 34 and 38.22%, respectively, when compared with those of corresponding controlled fruit samples.

Keywords Bionanocomposites · Overall migration rate · Migration rate of nanoparticles · Food packaging · Shelf life

When nanocomposite films start to be in contact with foodstuffs, the migration rate of constituents in bionanocomposite films needs to be taken into account. The migration process can be simply defined as the mass transfer of low-molecular-weight molecules from packaging materials to packed products like foodstuffs [1, 2]. The migrated molecules may be plasticisers, nanofillers and other additives like surfactants [1, 2]. Furthermore, the migration rates of these molecules are deemed as a function of their molecular weight, concentration, solubility and diffusivity, as well as other conditions like pH level, temperature and contact time between packaging materials and packed products [1, 3]. Migration process can be considered as a diffusion process according to Fick's second law [4]. The European Commission Regulation (EU) No. 10/2011 [5] specifies the safety limit for the migrated molecules should not exceed the OML of 60 mg/kg onto foodstuffs, which is equivalent to 10 mg/dm^2 for packaging materials. Moreover, the European Commission Regulation (EU) No. 10/2011 [5] clearly indicates the limits for some elements like barium, copper, cobalt, iron, lithium, manganese and zinc. Consequently, the migration process in terms of overall migration rate and HNT

© Springer Nature Singapore Pte Ltd. 2020
Z. W. Abdullah and Y. Dong, *Polyvinyl Alcohol/Halloysite Nanotube Bionanocomposites as Biodegradable Packaging Materials*,
https://doi.org/10.1007/978-981-15-7356-9_6

migration rate have been covered in this chapter, and their results can be implemented to present the validity of bionanocomposite films as food packaging materials.

6.1 Overall Migration Rate

The overall migration rates of PVA/ST/GL blend and its corresponding bionanocomposite films at different HNT contents were evaluated in three food simulants, as illustrated in Fig. 6.1. These food simulants, including a 10% ethanol solution (simulant A), a 3% acidic acid solution (simulant B) at the pH level below 4.5 and a 50% ethanol solution (simulant D1), were selected to mimic hydrophilic, acidic and lipophilic foodstuffs, respectively. It can be clearly seen from Fig. 6.1 that PVA/ST/GL blends and their bionanocomposite films have higher migration rates in food simulant A as opposed to other food simulants due to the hydrophilic nature of film constituents. The overall migration rates of these films exceed the OML of 60 mg/kg [5] in food simulant A except bionanocomposite films reinforced with 1 and 3 wt% HNTs. This behaviour may be interpreted depending on water solubility characteristics of blend matrices and their bionanocomposite films. In other words, PVA and ST are water-soluble polymers and their blends have high solubility rates in hydrophilic media such as food simulant A. Furthermore, this solubility of polymer blends diminishes in the presence of hydrophobic nanofillers like HNTs, which is reflected by the reduction in the overall migration rate of bionanocomposite films at different HNT contents in comparison with those of polymer blends (see Chap. 4). Moreover, strong interfacial bonding between blend matrices and HNTs is another reason to reduce mass transfer when these films contact hydrophilic media. Consequently, remarkable reductions in overall migration rates by 47.41 and 45.85% have been reported for PAV/ST/GL/HNT bionanocomposite films at the HNT contents of 1 and 3 wt%, respectively, as opposed to that of blend matrices in food simulant A. When HNT contents are increased up to 5 wt%, the overall migration rate of bionanocomposite films is enhanced by 30.97% as well due to HNT agglomeration. However, it is still lower than those of blend matrices and bionanocomposite films reinforced with 0.25 and 0.5 wt% HNTs. Similar results have also been obtained by Cano et al. [6] based on PVA/ST/ cellulose nanocrystal (CNC) nanocomposite films.

A completely different trend has been observed in food simulant D1. The overall migration rates of PVA/ST/GL blends and their bionanocomposite films are lower than the OML of 60 mg/kg. A clear reduction in overall migration rate has been found for bionanocomposite films at the HNT contents in range of 0.25–3 wt% when compared with that of corresponding blends. In particular, the maximum reduction has been found to be 19.93% with the inclusion of 0.25 wt% HNTs in bionanocomposites.

The lowest overall migration rates of PVA/ST/GL blends and their bionanocomposite films have been reported in food simulant B. Such a trend reflects

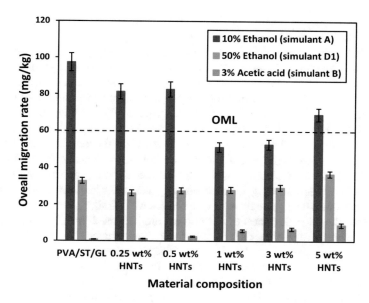

Fig. 6.1 Overall migration rates of PVA/ST/GL blend and its corresponding bionanocomposite films at various HNT contents in three different food simulants [7]

the limited interactions between material films and acidic media like food simulant B. The overall migration rates are increased in a linear manner from 1.29 to 9.13 mg/kg with increasing the HNT content from 0 to 5 wt%. As a whole, the overall migration rates of blends and corresponding bionanocomposite films highly depend on the selection of food simulants. In other words, these films could be highly affected by hydrophilic foodstuffs, which is followed by lipidic foodstuffs along with the least impact by acidic foodstuffs. Such overall lower rates of migration in lipidic and acidic foodstuffs could be associated with the good resistance of PVA to oil, grease and solvents relative to the lower resistance of water.

6.2 HNT Migration Rate

The presence of Al^+ and Si^+ in migrated molecules has been used to evaluate the migration rate of HNTs because of their chemical structure of $Al_2Si_2O_5(OH)_4 \cdot nH_2O$ [8]. Consequently, the migration rates of Al^+ and Si^+ from PVA/ST/GL blends and their bionanocomposite films at different HNT contents are depicted in Fig. 6.2a, b, respectively. The migrated molecules from blend matrices do not show any traces of Al^+ and Si^+, which means that the migrated molecules shown in the overall migration rate results completely reflect the quantities of migrated polymeric molecules in three different food simulants. The migration rates of Al^+ and Si^+

based on bionanocomposite films agree well with the overall migration rates in different food simulants (see Fig. 6.1). That is to say, the higher quantities of Al^+ and Si^+ have been detected in food simulant A, followed by food simulant D1 and finally food simulant B. This trend could be interpreted based on hydrophilic nature of food simulant A, which works as a plasticiser and a solvent concurrently leading to the improvement of chain mobility of polymeric molecules and easy release of nanofillers from bionanocomposite films, respectively. Similar behaviour was observed by Lee et al. [9] based on chitosan/clove essential oil (CEO)/HNT nanocomposite films. Their results have showed that the release rate of CEO in lipidic food simulant is faster than those of hydrophilic and acidic food simulants in that food simulant molecules with oily nature diffuse into the structures of nanocomposite films leading to weak bonding networks between constituents and easy release of active agents from film structures with similar characteristics. On the other hand, the migration rates of Al^+ and Si^+ are increased linearly with increasing the HNT content regardless of food simulants. For instance, the migration rates of Al^+ and Si^+ are increased by 766.67 and 424.82%, respectively, with increasing the HNT content from 0 to 5 wt% in food simulant A. A similar increasing trend has been reported in food simulant D1 by 381.82 and 202.87% for Al^+ and Si^+, as well as 162.79 and 290.38% in food simulant B, respectively. It is evidently shown that the lower quantities of migrated Al^+ and Si^+ are revealed in food simulant B due to weak interactions between bionanocomposite films and acidic media, which is in good accordance with the overall migration results mentioned earlier based on the dual sorption theory [10, 11]. Because the *diffusion* process of food simulant molecules within bionanocomposite films takes place at the faster pace than *embedding* process of intermolecular forces between penetrant molecules and bionanocomposite films, both intermolecular spaces of polymeric chains and the release of HNTs from bionanocomposite films are improved accordingly. Apparently, the diffusion of food simulant molecules within material films plays an important role in controlling the migration rates. In other words, the diffusion process of food simulant molecules would be faster when the material films and food simulants have the same characteristics (for example, both are hydrophilic materials) leading to increasing the release rate of active agents from the film materials.

Unfortunately, there are no specific migration limits for Al^+ and Si^+ determined by the European Union Commission Regulation (EU) No 10/2011 [5] for a comparison purpose. However, HNTs as natural nanofillers are generally classified as a nontoxic, biocompatible and EPA-4A material [12–14]. Consequently, HNTs are well recognised as a good nanofiller candidate for several biomedical applications such as drug delivery for non-injectable drug formula, dentist resin, bone cement, tissue scaffolds and cosmetics [12, 14, 15]. The toxicity of HNTs has been studied over decades for *in vitro* and *in vivo* tests. Some of these studies have showed that HNTs did not possess the toxicity up to 75 µg/mL, while 90% of tested cells were still viable [15–17]. This safe concentration of HNTs has also been extended to 100 µg/mL based on other investigations [14], and then further increased up to 200 µg/mL [18]. When comparing the migrated rates of HNTs from PVA/ST/GL/ HNT bionanocomposite films with these safe concentrations (based on part per

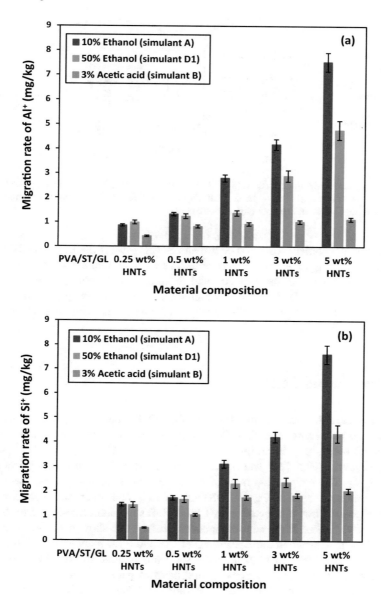

Fig. 6.2 Migration rate of **a** Al⁺ and **b** Si⁺ from PVA/ST/GL blend and its bionanocomposite films at various HNT contents in three different food simulants [7]

million as a comparison scale), these migrated rates still remain within the safe limits without any toxic effect on human bodies.

It is clearly demonstrated from Fig. 6.2a, b that there are no great differences between the migrated quantities of Al⁺ and Si⁺ for each food simulant type due to

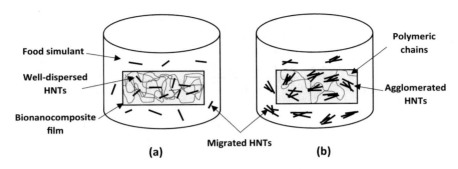

Fig. 6.3 Schematic diagram of released HNTs from PVA/ST/GL/HNT bionanocomposite films in a typical food simulant based on **a** well-dispersed HNTs at low nanofiller contents and **b** agglomerated HNTs at high nanofiller contents [7]

similar element contents of Al (20.90%) and Si (21.76%) in as-received HNTs [19]. Moreover, the agglomeration issue of HNTs at their content of 5 wt% clearly increases the migration rates of Al^+ and Si^+ due to the poor interfacial bonding between HNTs and blend matrices, which improves the releasing process of HNTs at high content levels from their films when compared with that at low HNT contents (see Fig. 6.3).

6.3   Packaging Tests

Based on migration rate results, PVA/ST/GL/HNT bionanocomposite films could be used as packaging materials for lipidic and acidic foodstuffs. Consequently, freshly cut avocados with a lipid content of 20% ($\approx$18.7/100 g) [20] and peaches with pH level ≤ 3.5 [21] were selected to mimic lipidic and acidic foodstuffs, respectively. The weight loss rates of controlled avocados and peaches, as well as packed fruits in neat PVA, PVA/ST/GL blends and their corresponding bio-nanocomposite films reinforced with 1 wt% HNTs are summarised in Fig. 6.4a, b, respectively. Packed fruits with bionanocomposite films show much lower rates of weight loss as opposed to those using neat PVA and PVA blends due to lower water and gas permeabilities of bionanocomposite films relative to neat PVA and blend counterparts (see Chap. 5). Avocados and peaches packed with bionanocomposite films possess the weight losses of 25.24 and 18.05%, respectively, in contrast to 35.15 and 25.93% for those based on PVA/ST/GL blends. Loryuenyong et al. [22] concluded the similar results from less fruit contacts with air and oxygen due to enhanced barrier properties leading to the inhibition of the production and action of ethylene, which plays an important role during the fruit ripening process. Moreover, fruits packed with neat PVA show lower weight loss rate than that of PVA/ST/GL blends due to the former's lower gas permeability. Nonetheless, the weight loss rate

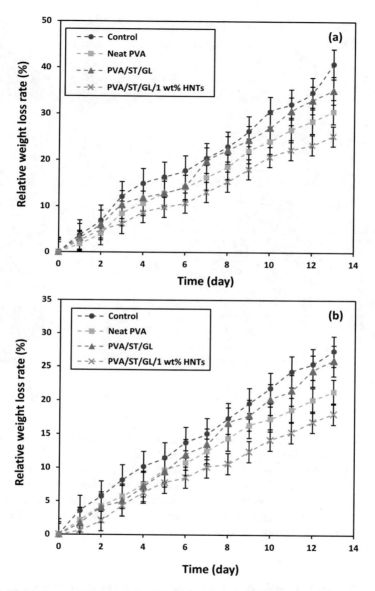

Fig. 6.4 Weight losses of **a** avocados and **b** peaches versus storage time for different packaging materials. Figures taken from [23] with copyright permission from Elsevier

of fruits packed with PVA/ST/GL blends appears to be still better than those of controlled samples.

Moreover, fungi have been found to grow on the external surfaces of controlled avocado and peach samples due to suitable environmental conditions of oxygen

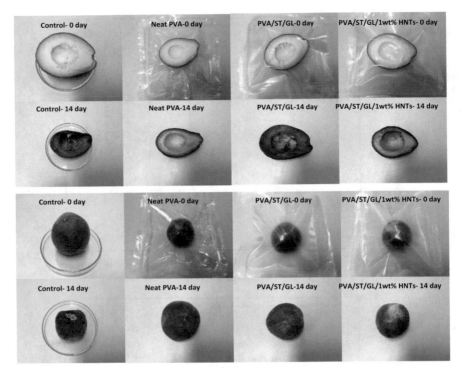

Fig. 6.5 External appearance of **a** avocados and **b** peaches before and after packaging tests with different packaging materials. Images taken from [23] with copyright permission from Elsevier

availability and humidity level, while lower gas and water permeabilities are main reasons to restrict the fungi growth on packed fruits, as illustrated in Fig. 6.5. On the other hand, fruits packed with bionanocomposite films give rise to lower colour changes with the extended shelf life in contrast with those using PVA/ST/GL blends. Accordingly, the improvement of barrier properties of PVA/ST/GL/HNT bionanocomposite films appears to be more pronounced for the extension of fruit shelf life relative to that of PVA/ST/GL blends.

6.4 Potential Applications as Food Packaging Materials

Petroleum-based polymers replaced other materials like metals, ceramics and wood in a wide variety of applications such as appliances, constructions and material packaging about half a century ago [24, 25]. These polymers such as polyethylene (PE), polystyrene (PS), polyamide (PA), polyvinyl chloride (PVC) and polypropylene (PP) become indispensable in most industrial sectors due to their reasonable mechanical, thermal and barrier properties, as well as low cost, relatively lightweight feature and excellent processability [4, 26]. According to the global market for polymeric consumption, about 322 million tons of polymers were consumed in 2015, which was increased by 3.5% as compared with that in 2014 [27]. Material packaging applications are on the top place at 42% of this consumption, particularly for food packaging in form of sheets, films, cups, trays and bottles due to the shift from reusable to single-use products [28, 29]. Moreover, 79% of this annual consumption is accumulated in the natural environment as plastic wastes since only 9% of this consumption is recycle, while 12% is burned leading to the significant increase in global warming [30]. Together, the lack of petroleum-based resources and their high cost of extraction, as well as their shortage in the next 60 years motivate the researchers in both academia and industrial sectors to find alternative and ecofriendly polymeric resources [28, 31].

To eliminate environmental plastic wastes, new ecofriendly material systems should be explored and developed to target commercial applications particularly for food packaging. In general, food packaging materials are required to possess good mechanical, thermal, optical and barrier properties in order to keep food quality in handling and storage processes, as well as biodegrade over a relatively short period of time to reduce the plastic waste issue [32, 33]. In the meantime, a good balance should be made for such materials between the aspects of material performance, cost, environment and human health through the selection of available raw materials, processing methods and interactions with food products as well as human beings [28, 34]. Existing systems for food packaging applications have relied on polymer/clay bionanocomposites since 1990 due to their excellent mechanical, thermal and barrier properties [34, 35]. Most of previous studies were focused on material preparation and characterisation with very little attention drawn to their applications as food packaging materials. Consequently, it is crucial that the application of bionanocomposite films as food packaging material is targeted in this study apart from their preparation and characterisation methods. Accordingly, it has been concluded that PVA/ST/GL/HNT bionanocomposite films can be successfully employed for packaging acidic and lipidic foodstuffs.

6.5 Summary

The migration of material constituents to the foodstuffs should be evaluated when considered as food packaging materials. Three different food simulant solutions of 10% ethanol (simulant A), 50% ethanol (simulant D1) and 3% acidic acid (simulant B) were used to study the overall migration rates, as well as the migration rates of HNTs. For food simulant A, the overall migration rates of PVA/ST/GL blends and their bionanocomposite films, except those with the inclusion of 1 and 3 wt% HNTs, exceed the OML of 60 mg/kg due to the hydrophilic nature of food simulant A with better solubility of films. Whereas, the lower overall migration rates of PVA/ST/GL blends and corresponding bionanocomposite films have been recorded in food simulant B though these migration rates are increased with increasing the HNT content, possibly resulting from poor interactions between material films and acidic foodstuffs. Moreover, the presence of HNTs reduces the overall migration rates of bionanocomposite films when compared with PVA/ST/GL blends in food simulant D1 despite being slightly increased at the HNT content of 5 wt%.

The quantities of Al^+ and Si^+ detected in migrated molecules are considered as an indicator for the migration rates of HNTs. The migration rates of HNTs are increased linearly with increasing the HNT content in three different food simulants due to the agglomeration issue of HNTs. This finding arises from weak interfacial bonding between blend matrices and nanofillers to assist in the release of HNTs from blend matrices. The migration rates of Al^+ and Si^+ have a similar trend to overall migration rates. In other words, the higher migration rates of Al^+ and Si^+ have been reported in food simulant A resulting from the solubility of material films in hydrophilic media, which is followed by food simulant D1 and then food simulant B due to their limited interactions with material films.

Based on migration test results, PVA/ST/GL/HNT bionanocomposite films are more applicable as the packaging materials for lipidic and acidic foodstuffs. Neat PVA, PVA/ST/GL blend and bionanocomposite films reinforced with 1 wt% HNTs can be used to pack freshly cut avocados and peaches to mimic lipidic and acidic foodstuffs, respectively. The weight loss rates of packed fruits with bio-nanocomposite films appear to be lower than those of controlled fruits, and fruits packed with neat PVA and PVA/ST/GL blends. This phenomenon could be interpreted by the improved barrier properties of bionanocomposite films with the incorporation of HNTs, as opposed to those of neat PVA and PVA/ST/GL blends, in order to restrict the effect of ethylene in the ripening process. Moreover, there is no noticeable fungi growth on the packed fruit surfaces when compared with controlled fruit samples despite the change of colours, which clearly indicates the improvement of fruit shelf life.

References

1. Huang JY, Li X, Zhou W (2015) Safety assessment of nanocomposite for food packaging application. Trend Food Sci Technol 45(2):187–199
2. Arvanitoyannis IS, Bosnea L (2004) Migration of substances from food packaging materials to foods. Crit Rev Food Sci Nutr 44(2):63–76
3. Avella M, Vlieger JJD, Errico ME, Fischer S, Vacca P, Volpe MG (2005) Biodegradable starch/clay nanocomposite films for food packaging applications. Food Chemist 93(3):467–474
4. Souza VGL, Fernando AL (2016) Nanoparticles in food packaging: biodegradability and potential migration to food—a review. Food Packag Shelf Life 8:63–70
5. European Union Commission regulation (EU) No 10/2011 (2011) On plastic materials and articles intended to come into contact with food. Offic J Europ Union L12:1–89
6. Cano A, Fortunati E, Chafer M, Gonzalez-Martinez C, Chiralt A, Kenny JM (2015) Effect of cellulose nanocrystals on the properties of pea starch–poly(vinyl alcohol) blend films. J Mater Sci 50(21):6979–6992

7. Abdullah ZW, Dong Yu (2019) Biodegradable and water resistant poly(vinyl) alcohol (PVA)/ starch (ST)/glycerol (GL)/halloysite nanotube (HNT) nanocomposite films for sustainable food packaging. Frontiers Mater 6:58

8. Khoo WS, Ismail H, Ariffin A (2011) Tensile and swelling properties of polyvinyl alcohol/ chitosan/halloysite nanotubes nanocomposite. Paper presented at the national postgraduate conference, IEEE, Kuala Lumpur, Malaysia, 19–20 Sept 2011

9. Lee MH, Kim SY, Park HJ (2018) Effect of halloysite nanoclay on the physical, mechanical, and antioxidant properties of chitosan films incorporated with clove essential oil. Food Hydrocoll 84:58–67

10. Farhoodi M, Mousavi SM, Sotudeh-Gharebagh R, Emam-Djomeh Z, Oromiehie A (2014) Migration of aluminum and silicon from PET/clay nanocomposite bottles into acidic food simulant. Packag Technol Sci 27(2):161–168

11. Huang Y, Chen S, Bing X, Gao C, Wang T, Yuan B (2011) Nanosilver migrated into food-simulating solutions from commercially available food fresh containers. Packag Technol Sci 24(5):291–297

12. Kamble R, Ghag M, Gaikawad S, Panda BK (2012) Halloysite nanotubes and applications: a review. J Adv Sci Res 3(2):25–29

13. Kryuchkova M, Danilushkina A, Lvov Y, Fakhrullin R (2016) Evaluation of toxicity of nanoclays and graphene oxide in vivo: a paramecium caudatum study. Environ Sci Nano 3 (2):442–452

14. Lvov YM, DeVilliers MM, Fakhrullin RF (2016) The application of halloysite tubule nanoclay in drug delivery. Expert Opin Drug Deliv 13(7):977–986

15. Fizir M, Dramou P, Dahiru NS, Ruya W, Huang T, He H (2018) Halloysite nanotubes in analytical sciences and in drug delivery: a review. Mikrochim Acta 185(8):389

16. Santos AC, Ferreira C, Veiga F, Ribeiro AJ, Panchal A, Lvov Y, Agarwal A (2018) Halloysite clay nanotubes for life sciences applications: from drug encapsulation to bioscaffold. Adv Colloid Interf Sci 257:58–70

17. Vergaro V, Abdullayev E, Lvov YM, Rinaldi R, Zeitoun A, Leporatti S (2010) Cytocompatibility and uptake of halloysite clay nanotubes. Biomacromol 11:820–826

18. Guo M, Wang A, Muhammad F, Qi W, Ren H, Guo Y, Zhu G (2012) Halloysite nanotubes, a multifunctional nanovehicle for anticancer drug delivery. Chinese J Chemist 30(9):2115–2120

19. Mousa MH, Dong Y, Davies IJ (2016) Recent advances in bionanocomposites: preparation, properties, and applications. Inter J Polymer Mater Polymer Biomater 65(5):225–254

20. Seymour GB, Tucker GA (1993) Avocado. In: Seymour GB, Taylor JE, Tucker GA (eds) Biochetnistry of fruit ripening. Springer, Malaysia, pp 53–82

21. Brady CJ (1993) Stone fruit. In: Seymour GB, Taylor JE, Tucker GA (eds) Biochetnistry of fruit ripening. Springer, Malaysia, pp 379–397

22. Loryuenyong V, Saewong C, Aranchaiya C, Buasri A (2015) The improvement in mechanical and barrier properties of poly(vinyl alcohol)/graphene oxide packaging films. Packag Technol Sci 28(11):939–947

23. Abdullah ZW, Dong Y, Han N, Liu S (2019) Water and gas barrier properties of polyvinyl alcohol (PVA)/starch (ST)/ glycerol (GL)/halloysite nanotube (HNT) bionanocomposite films: experimental characterisation and modelling approach. Compos B Eng 174:107033

24. Shah AA, Hasan F, Hameed A, Ahmed S (2008) Biological degradation of plastics: a comprehensive review. Biotechnol Adv 26(3):246–265

25. Arora A, Padua GW (2010) Review: nanocomposites in food packaging. J Food Sci 75(1):43–49

26. Siracusa V, Rocculi P, Romani S, Rosa MD (2008) Biodegradable polymers for food packaging: a review. Trend Food Sci Technol 19:634–643

27. Mangaraj S, Yadav A, Bal LM, Dash SK, Mahanti NK (2018) Application of biodegradable polymers in food packaging industry: a comprehensive review. J Packag Technol Res 3 (1):77–96

28. Silvestre C, Duraccio D, Cimmino S (2011) Food packaging based on polymer nanomaterials. Prog Polym Sci 36(12):1766–1782
29. Ramos ÓL, Pereira RN, Cerqueira MA, Martins JR, Teixeira JA, Malcata FX, Vicente AA (2018) Bio-based nanocomposites for food packaging and their effect in food quality and safety. In: Holban AM, Grumezescu AM (eds) Food packaging and preservation. Elsevier, London, UK, pp 271–306
30. Geyer R, Jambeck JR, Law KL (2017) Plastics: production, use, and fate of all plastics ever made. Sci Adv 3:1700782
31. Mishra RK, Ha SK, Verma K, Tiwari SK (2018) Recent progress in selected bio-nanomaterials and their engineering applications: an overview. J Sci Adv Mater Devic 3(3):263–288
32. Rhim JW, Park HM, Ha CS (2013) Bio-nanocomposites for food packaging applications. Prog Polym Sci 38:1629–1652
33. Rhim JW, Ng PKW (2007) Natural biopolymer-based nanocomposite films for packaging applications. Crit Rev Food Sci Nutr 47(4):411–433
34. Ray S, Quek SY, Easteal A, Chen XD (2006) The potential use of polymer-clay nanocomposites in food packaging. Int J Food Eng 2(4):5
35. Majeed K, Jawaid M, Hassan A, Abu-Bakar A, Abdul Khalil HPS, Salema AA, Inuwa I (2013) Potential materials for food packaging from nanoclay/natural fibres filled hybrid composites. Mater Desig 46:391–410

Printed in the United States
by Baker & Taylor Publisher Services